SPICE for Circuits and Electronics Using PSpice®

MUHAMMAD H. RASHID

Professor of Electrical Engineering
Purdue University at Fort Wayne

PRENTICE HALL

Englewood Cliffs, New Jersey   07632

Library of Congress Cataloging-in-Publication Data

Rashid, M. H.
   SPICE for circuits and electronics using PSpice / Muhammad H.
Rashid.
      p.   cm.
   Includes bibliographies and index.
   ISBN 0-13-834672-0
   1. Electric circuit analysis—Data processing.   I. Title.
TK454.R385   1990
621.319′2′0285—dc20   89-33565
                                                       CIP

Editorial/production supervision: *Edith Riker/DebraAnn Thompson*
Cover design: *Ben Santora*
Manufacturing buyer: *Robert Anderson*

PSpice and *Probe* are registered trademarks of MicroSim Corporation.
IBM-PC is a registered trademark of International Business Machines
Corporation.
Macintosh-II is a registered trademark of Apple Computer, Inc.
WordStar and *Wordstar* 2000 are the registered trademarks of Micropro
International Corporation.
Word is a registered trademark of Microsoft Corporation.
Program Editor is a registered trademark of WordPerfect Corporation.

 © 1990 by Prentice-Hall, Inc.
A Division of Simon & Schuster
Englewood Cliffs, New Jersey 07632

Printed in the United States of America

10 9 8 7 6 5 4

ISBN   0-13-834672-0

Prentice-Hall International (UK) Limited, *London*
Prentice-Hall of Australia Pty. Limited, *Sydney*
Prentice-Hall Canada, Inc., *Toronto*
Prentice-Hall Hispanoamericana, S.A., *Mexico*
Prentice-Hall of India Private Limited, *New Delhi*
Prentice-Hall of Japan, Inc., *Tokyo*
Simon & Schuster Asia Pte. Ltd., *Singapore*
Editora Prentice-Hall do Brasil, Ltda., *Rio de Janeiro*

To my parents
my wife—Fatema
and
children—Fa-eza, Farzana, Hasan

Contents

PREFACE   *xiii*

CHAPTER 1   INTRODUCTION   **1**

     1.1   Introduction   1

     1.2   Descriptions of SPICE   2

     1.3   Types of Spice   3

     1.4   Limitations of *PSpice*   3

           References   4

CHAPTER 2   CIRCUIT DESCRIPTIONS   **5**

     2.1   Introduction   5

     2.2   Nodes   6

     2.3   Element Values   6

2.4   Circuit Elements   7

2.5   Element Models   9

2.6   Sources   9

2.7   Types of Analysis   10

2.8   Output Variables   11

2.9   *PSpice* Output Commands   11

2.10   Format of Circuit Files   11

2.11   Format of Output Files   12

        References   18

        Problems   18

CHAPTER 3   DEFINING OUTPUT VARIABLES   **20**

3.1   Introduction   20

3.2   DC Sweep and Transient Analysis   20
        3.2.1   Voltage Output, 21
        3.2.2   Current Output, 22

3.3   AC Analysis   24
        3.3.1   Voltage Output, 24
        3.3.2   Current Output, 24

3.4   Noise Analysis   26

        Summary   26

CHAPTER 4   VOLTAGE AND CURRENT SOURCES   **28**

4.1   Introduction   28

4.2   Sources Modeling   28
        4.2.1   Exponential Source, 29
        4.2.2   Pulse Source, 30
        4.2.3   Piecewise Linear Source, 31
        4.2.4   Single-Frequency
        *   Frequency-Modulation, 31*
        4.2.5   Sinusoidal Source, 32
        4.2.6   Polynomial Source, 33

4.3   Independent Sources   36
        4.3.1   Independent Voltage Source, 36
        4.3.2   Independent Current Source, 37

4.4   Dependent Sources   37
        *4.4.1   Voltage-Controlled Voltage
                 Source, 37*
        *4.4.2   Voltage-Controlled Current
                 Source, 39*
        *4.4.3   Current-Controlled Current
                 Source, 40*
        *4.4.4   Current-Controlled Voltage
                 Source, 40*

        Summary   41

        Problems   42

CHAPTER 5   PASSIVE ELEMENTS   **45**

5.1   Introduction   45

5.2   Modeling of Elements   45

5.3   Operating Temperature   46

5.4   RLC-Elements   47
        5.4.1   Resistor, 47
        5.4.2   Capacitor, 48
        5.4.3   Inductor, 49

5.5   Magnetic Elements   52

5.6   Lossless Transmission Lines   56

5.7   Switches   57
        5.7.1   Voltage-Controlled Switch, 58
        5.7.2   Current-Controlled Switch, 60

        Summary   63

        References   63

        Problems   63

CHAPTER 6   DOT COMMANDS   **67**

6.1   Introduction   67

6.2   Model   68
        6.2.1   .MODEL Model, 68
        6.2.2   .SUBCKT Subcircuit, 68
        6.2.3   .ENDS End of Subcircuit, 68

6.2.4 .LIB Library File, 70
6.2.5 .INC Include File, 70

6.3   Types of Output   71
        6.3.1 .PRINT Print, 71
        6.3.2 .PLOT Plot, 71
        6.3.3 .PROBE Probe, 72
        6.3.4 Probe Output, 73
        6.3.5 .WIDTH Width, 77

6.4   Temperature and End of Circuit   77

6.5   Options   78

6.6   DC Analysis   81
        6.6.1 .OP Operating Point, 81
        6.6.2 .NODESET Nodeset, 82
        6.6.3 .SENS Sensitivity Analysis, 82
        6.6.4 .TF Small-Signal Transfer
               Function, 84
        6.6.5 .DC DC Sweep, 86

6.7   AC Analysis   88

6.8   Noise Analysis   89

6.9   Transient Response   91
        6.9.1 .IC Initial Transient
               Conditions, 92
        6.9.2 .TRAN Transient Analysis, 92

6.10   Fourier Analysis   95

        Summary   97

        Problems   98

CHAPTER 7   SEMICONDUCTOR DIODES   101

7.1   Introduction   101

7.2   Diode Model   101

7.3   Diode Statement   104

        Summary   116

        References   116

        Problems   116

CHAPTER 8   BIPOLAR JUNCTION TRANSISTORS   119

8.1   Introduction   119

8.2   BJT Model   119

8.3   BJT Statements   123

Summary   149

References   149

Problems   149

CHAPTER 9   FIELD-EFFECT TRANSISTORS   153

9.1   Introduction   153

9.2   Junction Field-Effect Transistors   153

9.3   Metal Oxide Silicon Field-Effect Transistors   165

9.4   Gallium Arsenide MESFETs   177

Summary   182

References   182

Problems   183

CHAPTER 10   OP-AMP CIRCUITS   186

10.1   Introduction   186

10.2   DC Linear Models   187

10.3   AC Linear Model   187

10.4   Nonlinear Macromodel   188

References   201

Problems   202

CHAPTER 11   DIFFICULTIES   206

11.1   Introduction   206

11.2   Large Circuits   207

11.3   Running Multiple Circuits   207

11.4   Large Outputs   207

11.5   Long Transient Runs   208

11.6   Convergence Problems   208
        11.6.1   DC Sweep, 209
        11.6.2   Bias Point, 211
        11.6.3   Transient Analysis, 212

11.7   Analysis Accuracy   212

11.8   Negative Component Values   212

11.9   Power Switching Circuits   213

11.10   Floating Nodes   216

11.11   Nodes With Less Than Two
        Connections   219

11.12   Voltage Source and Inductor
        Loops   220

11.13   Running PSpice File on Spice   221

11.14   Running Spice File on PSpice   221

        References   222

        Problems   222

APPENDIX A   RUNNING PSpice ON PCs   223

A.1   Creating Input Files   224

A.2   Run Command   224

A.3   DOS Commands   225

APPENDIX B   NOISE ANALYSIS   226

B.1   Thermal Noise   226

B.2   Shot Noise   227

B.3   Flicker Noise   227

B.4   Burst Noise   228

B.5   Avalanche Noise   228

B.6   Noise in Diodes   228

B.7   Noise in Bipolar Transistors, 229

B.8   Noise in Field-Effect Transistors   230

B.9   Equivalent Input Noise   231

APPENDIX C   NONLINEAR MAGNETIC   **234**

BIBLIOGRAPHY   **236**

INDEX   **237**

Notes:

The students are urged to print the contents of files "README.DOC" to set up for the right printer and monitor, and "NOM.LIB" to obtain the list of subcircuits for semiconductor devices. These files come with the student version of PSpice software programs and can be printed by the DOS TYPE command.

Preface

The Accreditation Board of Engineering and Technology (ABET) requirements specify the integration of computer-aided analysis and design in EE-curriculum. SPICE is a very popular software for analyzing electrical and electronic circuits. Until recently, a mainframe or a VAX-class computer was required. Beside the cost, such a machine is not convenient for assignments in classes of 200-level circuits and electronics courses. Recently, MicroSim Corporation introduced the *PSpice* simulator that can run on personal computers (PCs) and is similar to the University of California (UC) Berkeley SPICE. The student version of *PSpice,* which is available free to students, is ideal for classroom use and for assignments requiring computer aided simulation and analysis. *PSpice* widens the scope for the integration of computer aided simulation to circuits and electronics courses for undergraduate students.

SPICE simulation can be taught as a 1-credit-hour course. The EE-curriculum is constantly under pressure to add new courses to cope with the changing technology. As a result, the credit hours are being increased. It may not be possible to add a 1-credit-hour course on SPICE to integrate computer aided analysis in circuits and electronics courses. However, the students require some basic knowledge of using SPICE. They are constantly under pressure with course loads and do not always have free time to read the details of SPICE from manuals and books of general nature.

This book is the outcome of difficulties faced by the author in integrating SPICE in circuits and electronics courses at the 200-level. The objective of this book is to introduce the SPICE simulator to the EE-curriculum at the sophomore or junior level with a minimum amount of time and effort. This book requires no prior knowledge about the SPICE simulator. A course on basic circuits should be a prerequisite or co-requisite. Once the students develop interests and appreciations in the applications of circuit simulator like SPICE, they can read advanced materials for the full utilization of SPICE or *PSpice* in solving complex circuits and systems.

This book can be divided into five parts: (1) Introduction to SPICE simulation—chapters 1, 2, and 3; (2) Source and element modeling—chapters 4 and 5; (3) SPICE commands—chapter 6; (4) Semiconductor devices modeling—chapters 7, 8, 9 and 10; and (5) Difficulties—chapter 11. Chapters 7, 8 and 9 describe the simple equivalent circuits of transistors and op-amps, which are commonly used in analyzing electronic circuits. Although SPICE generates the parameters of complex transistor models, the analysis with a simple circuit model exposes the students to the mechanism of computation by SPICE .MODEL commands. This approach has the advantage that the students can compare the results, which are obtained in a classroom environment with the simple circuit models of devices, to those obtained by using complex SPICE models.

The commands, models and examples that are described for *PSpice* are also applicable to UC Berkeley SPICE with minor modifications. The changes for running a *PSpice* circuit file on SPICE and vice-versa are discussed in Chapter 11. The circuit files in this book are typed in uppercase so that the same file can be run on either the *PSpice* or the SPICE simulator.

Probe is a graphics post-processor and is very useful in plotting the results of simulation, especially with the capability of arithmetic operation it can be used to plot impedance, power, etcetera. Once the students have experience in programming on *PSpice,* they really appreciate the advantages of .Probe command. *Probe* is an option on *PSpice,* available with the student version. Running *Probe* does not require a math co-processor. The students can also get the normal printer output or printer plotting. The prints and plots are very helpful to the students in relating their theoretical understanding and making judgment on the merits of a circuit and its characteristics.

This book can be used as a text book on SPICE with a course on basic circuits being the prerequisite or co-requisite. It can also be a supplement to any standard text book on Basic Circuits or Electronics. In the case of second option, the following sequence is recommended for the integration of SPICE at Basic Circuits level,

1. Supplement to a Basic Circuits course with 3 hours of lectures (or equivalent Lab hours) and self study assignments from chapters 1 to 6. Staring from chapter 2, the students should work with PCs.

2. Continue as a supplement to an Electronics course with 2 hours of lectures (or equivalent Lab hours) and self study assignments from chapters 7 to 11.

For integrating SPICE at the electronics level, 3 hours of lectures (or equivalent Lab hours) are recommended on chapters 1 to 6. Chapters 7 to 11 could be left for self study assignments. From the author's experience in the class, it has been observed that after three lectures of 50 minutes duration, all students could solve assignments independently without any difficulty. The class could progress in a normal manner with one assignment per week on electronic circuits simulation and analysis with SPICE. Although the materials of this book have been tested in a basic circuits course for engineering students and in an electronics course for EE-students, the book is also recommended for EET-students.

Muhammad H. Rashid
Munster, Indiana

ACKNOWLEDGMENTS

I would like to thank the following reviewers for their comments and suggestions:

Frank H. Hielscher, Lehigh University
A. Zielinski, University of Victoria, Canada
Emil C. Neu, Stevens Institute of Technology

It has been a great pleasure working with the editor, Elizabeth Kaster. Finally, I would thank my family for their love, patience, and understanding.

PSpice SOFTWARE

The *PSpice* student version software is available from Prentice-Hall. To order the software, please see the card which is included in this book.
—*PSpice* student version disks (2) IBM PC compatable (83463-0)
—*PSpice* student version disk (1) MAC-II compatable (83462-2)
—*PSpice* student version disk (1) IBM PS/2 compatable (83464-8)

Any Comments and suggestions regarding this book are welcomed and should be sent to the author.

Dr. Muhammad H. Rashid
Professor of Electrical Engineering
Indiana University-Purdue University
at Fort Wayne
Fort Wayne, Indiana 46805-1499

1

Introduction

1.1 INTRODUCTION

Electronic circuit design requires accurate methods for evaluating circuit performance. Because of the enormous complexity of modern integrated circuits, computer-aided circuit analysis is essential and can provide information about circuit performance that is almost impossible to obtain with laboratory prototype measurements. Computer-aided analysis permits

1. Evaluating the effects of variations in elements, such as resistors, transistors, transformers, etc.,
2. The assessment of performance improvements or degradations,
3. Evaluating the effects of noise and signal distortion without the need of expensive measuring instruments,
4. Sensitivity analysis to determine the permissible bounds due to tolerances on each and every element value or parameter of active elements,
5. Fourier analysis without expensive wave analyzers,
6. Evaluating the effects of nonlinear elements on the circuit performance, and
7. Optimizing the design of electronic circuits in terms of circuit parameters.

SPICE is a general-purpose circuit program that simulates electronic circuits. SPICE can perform various analysis of electronic circuits, e.g., the operating (or quiescent) points of transistors, time-domain response, small-signal frequency response, etc. SPICE contains models for common circuit elements, active as well as passive, and it is capable of simulating most electronic circuits. It is a versatile program and is widely used both in industries and universities. The acronym SPICE stands for *Simulation Program with Integrated Circuit Emphasis*.

Until recently, SPICE was available only on mainframe computers. In addition to the initial cost of the computer system, such a machine can be expensive and inconvenient for classroom use. In 1984, MicroSim introduced the *PSpice* simulator, which is similar to the Berkeley SPICE and runs on an IBM-PC or compatible. It is available free of cost to students for classroom use. *PSpice,* therefore, widens the scope for the integration of computer-aided circuit analysis into electronic circuits course at an undergraduate level. Other versions of *PSpice* that will run on computers such as Macintosh-II, 386-based processor, VAX, SUN, and NEC are also available.

1.2 DESCRIPTIONS OF SPICE

PSpice is a member of the SPICE family of circuit simulators, all of which originate from the SPICE2 circuit simulator. Its development spans a period of about 30 years. During the mid-1960s, the program ECAP was developed at IBM [1]. Later ECAP served as the starting point for the development of program CANCER at the University of California (UC), Berkeley, in the late 1960s. Using CANCER as the basis, SPICE was developed at UC Berkeley in early 1970s. During the mid-1970s, SPICE2, which is an improved version of SPICE, was developed at UC Berkeley. The algorithms of SPICE2 are robust, powerful, and general in nature; SPICE2 has become a standard tool in the industry for circuit simulations. SPICE3, which is a variation of SPICE2, is designed especially to support computer-aided design (CAD) research program at UC Berkeley. Since the development of SPICE2 was supported using public funds, this software is in the public domain, which means that it may be used freely by the citizens of the United States.

As an industry standard, SPICE2 is referred to simply as SPICE. The input syntax for SPICE is a free-format style that does not require data to be entered in fixed column locations. SPICE assumes reasonable default values of circuit parameters that are not specified. In addition, it performs a considerable amount of error checking to ensure that the circuit has been entered correctly.

PSpice, which uses the same algorithms as SPICE2, is equally useful for simulating all types of circuits in a wide range of applications. A circuit is described by statements that are stored in a file called a *circuit file*. The circuit file is read by the simulator. Each statement is self-contained and independent of the others and has no interaction with other statements. SPICE (or *PSpice*) statements are easy to learn and use.

1.3 TYPES OF SPICE

The commercially supported versions of SPICE2 can be divided into two types: mainframe versions and PC-based versions.

Mainframe versions are
HSPICE from Meta-Software
RAD-SPICE from Meta-Software
IG-SPICE from A.B. Associates
I-SPICE from NCSS Time Sharing
Precise from Electronic Engineering Software
PSpice from MicroSim

HSPICE is designed for integrated circuit design with special model support. RAD-SPICE simulates circuits subjected to ionizing radiation. IG-SPICE and I-SPICE are designed for interactive circuit simulation with graphics output.

PC-based versions are

AllSpice from Acotech
IS-SPICE from Intusoft
Z-SPICE from Z-Tech
SPICE-Plus from Analog Design Tools
DSPICE from Daisy Systems
PSpice from MicroSim

1.4 LIMITATIONS OF *PSpice*

PSpice as a circuit simulator has the following limitations:

1. The student version of *PSpice* that is PC-based is restricted to circuits with 10 transistors only. However, the professional (or production) version can simulate a circuit with up to 200 bipolar transistors (or 150 MOSFETs, Metal Oxide Semiconductor Field–Effect Transistors).
2. The program is not interactive. That is, the circuit cannot be analyzed for various component values without editing the program statements.
3. *PSpice* does not support an iterative method of solution. If the elements of a circuit are specified, the output can be predicted. On the other hand, if the

output is specified, *PSpice* can not be used to synthesize the circuit elements.

4. The input impedance cannot be determined directly without running the graphic post-processor, *Probe*. The student version does not require a floating-point coprocessor for running *Probe*. However, the professional version does.

5. The PC version needs 512 kbytes of memory (RAM) to run.

6. Distortion analysis is not available.

7. The output impedance of a circuit cannot be printed or plotted directly.

8. The student version will run *with* or *without* the floating-point coprocessor (8087, 80287, or 80387). If the coprocessor is present, the program will run at full speed. Otherwise it will run 5 to 15 times slower. The professional version requires the coprocessor (it is not optional).

REFERENCES

1. R. W. Jensen and M. D. Liberman, *IBM Electronic Circuit Analysis Program and Applications*. Englewood Cliffs, N.J.: Prentice Hall Inc., 1968.

2. R. W. Jensen and L. P. McNamee, *Handbook of Circuit Analysis Languages and Techniques*. Englewood Cliffs, N.J.: Prentice Hall Inc., 1976.

3. *PSpice Manual,* Irvine, Calif.: MicroSim Corporation, 1988.

2

Circuit Descriptions

2.1 INTRODUCTION

PSpice is a general-purpose circuit program that can be applied to simulate and calculate the performance of electronic circuits. A circuit must be specified in terms of element names, element values, nodes, variable parameters, and sources. Let us consider the circuit in Figure 2.1 that is to be simulated for calculating and plotting the transient response from 0 to 3 ms with a time increment of 10 μs. We shall show (1) how to describe this circuit to *PSpice*, (2) how to specify the type of analysis to be performed, and (3) how to define the required output variables. The descriptions and analysis of a circuit require specifying the following:

Nodes
Element values
Circuit elements
Element models
Sources
Types of analysis
Output variables

PSpice output commands
Format of circuit files
Format of output files

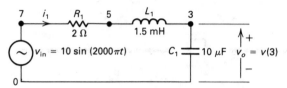

Figure 2.1   An RLC circuit

2.2 NODES

Node numbers are assigned to the circuit in Figure 2.1. Elements are connected
between nodes. The node numbers to which an element is connected are specified
after the name of the element. Node numbers must be integers from 0 to 9999.
The node numbers need not be sequential. Node 0 is predefined as the ground.
All nodes must be connected to at least two elements and should, therefore,
appear at least twice.

All nodes must have a DC path to the ground node. This condition, which is
not always satisfied in some circuits, is normally met by connecting very large
resistors and is discussed in Section 11.10.

2.3 ELEMENT VALUES

The value of a circuit element is written after the nodes to which the element is
connected. The values are written in standard floating-point notation with optional scale and units suffixes. Some values without suffixes that are allowable by
PSpice are

$$5 \qquad 5. \qquad 5.0 \qquad 5E + 3 \qquad 5.0E + 3 \qquad 5.E3$$

There are two types of suffixes: scale suffix and units' suffix. The scale suffix
multiplies the number that it follows. The scale suffixes recognized by *PSpice* are

$$F = 1E\text{-}15$$

$$P = 1E\text{-}12$$

$$N = 1E\text{-}9$$

$$U = 1E\text{-}6$$

$$MIL = 25.4E\text{-}6$$

$$M = 1E\text{-}3$$

$$K = 1E3$$

$$MEG = 1E6$$

$$G = 1E9$$

$$T = 1E12$$

The units' suffixes that are normally used are

$$V = \text{volt}$$

$$A = \text{amp}$$

$$HZ = \text{hertz}$$

$$OHM = \text{ohm } (\Omega)$$

$$H = \text{henry}$$

$$F = \text{farad}$$

$$DEG = \text{degree}$$

The first suffix is always the scale suffix, and the units' suffix follows the scale suffix. In the absence of a scale suffix, the first suffix may be a units suffix, provided it is not the symbol of scale suffixes. The units' suffixes are always ignored by *PSpice*. If the value of an inductor is 15 μH it is written as 15U or 15UH. In the absence of scale and units suffixes, the units of voltage, current, frequency, inductance, capacitance, and angle are, by default, volts, amps, hertz, henrys, farads, and degrees. *PSpice* ignores any units suffix, so the following values are equivalent

25E-3   25.0E-3   25M   25MA   25MV   25MOHM   25MH

Notes

1. The scale suffixes are all uppercase.
2. M means *milli*, not *mega*. 2MΩ is written as 2MEG or 2MEGOHM.

2.4 CIRCUIT ELEMENTS

Circuit elements are identified by names. A name must start with a letter symbol corresponding to the element, but after that it can contain either letters or numbers. Names can be up to eight characters long. Table 2.1 shows the first letter of elements and sources. For example, the name of a capacitor must start with *C*.

TABLE 2.1   SYMBOLS OF CIRCUIT ELEMENTS
AND SOURCES

First letter	Circuit Elements and Sources
B	GaAs MES field-effect transistor
C	Capacitor
D	Diode
E	Voltage-controlled voltage source
F	Current-controlled current source
G	Voltage-controlled current source
H	Current-controlled voltage source
I	Independent current source
J	Junction field-effect transistor
K	Mutual inductors (transformer)
L	Inductor
M	MOS field-effect transistor
Q	Bipolar junction transistor
R	Resistor
S	Voltage-controlled switch
T	Transmission line
V	Independent voltage source
W	Current-controlled switch

Note: Voltage-controlled switches and current-controlled switches are not available in SPICE2, but they are available in SPICE3.

The format for describing passive elements is

⟨element name⟩   ⟨positive node⟩   ⟨negative node⟩   ⟨value⟩

where the current is assumed to flow from a positive node, N+ to a negative node, N−. The formats for active and passive elements are described in Chapters 5, 7, 8, 9, and 10.

The passive elements in Figure 2.1 are described as follows:

- The statement for R1 that has a value of 2 Ω and is connected between nodes 7 and 5 is

```
R1    7    5    2
```

- The statement for L1 that has a value of 1.5 mH and is connected between nodes 5 and 3 is

```
L1    5    3    1.5MH
```

- The statement for C1 that has a value of 10 μF and is connected between nodes 3 and 0 is

```
Cl    3    0    10UF
```

2.5 ELEMENT MODELS

The values of some circuit elements are dependent on other parameters, such as the initial condition of an inductor, the capacitance as a function of voltage, and the resistance as a function of temperature. Models may be used to assign values to the various parameters of circuit elements. The techniques for specifying models of sources, passive elements, and active elements are described in Chapters 4, 5, and 6.

The model for a simple sinusoidal source is

```
SIN (VO    VA    FREQ)
```

where

> VO   = offset voltage, V
> VA   = peak voltage, V
> FREQ = frequency, Hz

For a sinusoidal voltage $v_{in} = 10 \sin(2\pi \times 1000t)$, the model is

```
SIN (0    10    1KHZ)
```

2.6 SOURCES

Voltage (or current) sources can be dependent or independent. The letter symbols for the names of sources are also listed in Table 2.1. An independent voltage (or current) source can be DC, sinusoidal, pulse, exponential, polynomial, piecewise linear, or single-frequency frequency-modulation. The models for describing the parameters of sources are discussed in Chapter 4.

The format for sources is

```
⟨source name⟩ ⟨positive node⟩ ⟨negative node⟩ ⟨source model⟩
```

where the current is assumed to flow into the source from a positive node, N+ to a negative node, N−. The order of nodes N+ and N− is critical. Assuming node 7 has a higher potential with respect to node 0, the statement for the input source, VIN, that is connected between nodes 7 and 0 is

```
VIN  7  0   SIN (0  10  1KHZ)
```

2.7 TYPES OF ANALYSIS

PSpice allows various types of analysis. Each analysis is invoked by including its command statement. For example, a statement beginning with a .DC command will cause the DC sweep to be done. The types of analysis and their corresponding .(dot) commands are as follows.

DC ANALYSIS

DC sweep of an input voltage/current source, a model parameter, or temperature (.DC)
Linearized device model parameterization (.OP)
DC operating point (.OP)
Small-signal transfer function (Thevenin's equivalent) (.TF)
Small-signal sensitivities (.SENS)

TRANSIENT ANALYSIS

Time domain response (.TRAN)
Fourier analysis (.FOUR)

AC ANALYSIS

Small-signal frequency response (.AC)
Noise analysis (.NOISE)

It should be noted that the . (*dot*) is an integral part of the commands. The various dot commands are discussed in detail in Chapter 6.
The format for performing transient response is

```
.TRAN  TSTEP  TSTOP
```

where

$\qquad$ TSTEP = time increment

$\qquad$ TSTOP = final (stop) time

Therefore, the statement for the transient response from 0 to 3 ms with a 10-μs increment is

```
.TRAN  10US  3MS
```

2.8 OUTPUT VARIABLES

PSpice has some unique features in printing or plotting output voltages or currents. The various types of output variables that are permitted by *PSpice* are discussed in Chapter 3.

The voltage of node 3 with respect to node 0 is specified by V(3, 0) or V(3).
The voltage of node 7 with respect to node 0 is specified by V(7, 0) or V(7).

2.9 *PSpice* OUTPUT COMMANDS

The most common forms of output are print tables and plots. The DC sweep (.DC), frequency response (.AC), noise (.NOISE), and transient response (.TRAN) analysis can produce output in the form of print tables and plots. The command for output in the form of tables is .PRINT, that for output plots is .PLOT, and that for graphical output is .PROBE.

The statement for the plots of V(3) and V(7) from the results of transient analysis is

```
.PLOT   TRAN   V(3)   V(7)
```

The statement for the tables of V(3) and V(7) from the results of transient analysis is

```
.PRINT   TRAN   V(3)   V(7)
```

Probe is a *graphic post-processor* for *PSpice;* the statement for the .PROBE command is

```
.PROBE
```

and the results of simulation are available for graphical outputs on the display and on the hard copy. After executing the .PROBE command, *Probe* will put up a menu on the screen to obtain graphical output. It is very easy to use *Probe*. The output commands are discussed in Section 6.3.

2.10 FORMAT OF CIRCUIT FILES

A circuit file that can be read by *PSpice* may be divided into five parts: (1) the title that describes the type of circuit or any comments, (2) the circuit description that defines the circuit elements and the set of model parameters, (3) the analysis description that defines the type of analysis, (4) the output description that defines the way the output is to be presented, and (5) the end of program by the .END command. The format for a circuit file is as follows:

    Title
    Circuit description
    Analysis description
    Output description
    .END (end of file statement)

Notes

1. The first line, which is the title line, may contain any type of text.
2. The last line must be the .END command.
3. The order of remaining lines is not important and does not affect the results of simulations.
4. If a *PSpice* statement is more than one line, the statement can continue in the next line. A continuation line is identified by a plus sign (+) in the first column of next line. The continuation lines must follow one another in the proper order.
5. A comment line may be included anywhere, with an asterisk (*) before comments.
6. The number of blanks between items is not significant (except the title line). The tabs and commas are equivalent to blanks. For example, " " and "   " and "," and "   ,   " are all equivalent.
7. *PSpice* statements or comments can be either uppercase or lowercase.
8. SPICE2 statements must be uppercase only. It is advisable to type the *PSpice* statements in uppercase so that the same circuit file can also be run on SPICE2.
9. If you are not sure of a command or statement, the best thing to do is to run the circuit file by using that command or statement to see what happens.
10. In electrical circuits, subscripts are normally assigned to symbols for voltages, currents, and circuit elements. However in *PSpice* the symbols are described without subscripts. For examples, v_i, i_1, R_1, L_1, C_1, V_{cc}, are represented by VI, I1, R1, L1, C1, and VCC, respectively. As a result, the *PSpice* circuit description of voltages and currents, and circuit elements can be different from those of circuit symbols.

11 FORMAT OF OUTPUT FILES

The results of simulation by *PSpice* are stored in an output file. It is possible to control the type and amount of output by various commands. If there is any error in the circuit file, *PSpice* will display a message on the screen indicating that there

is an error and will suggest looking at the output file for details. The output falls
into four types as follows:

1. Description of the circuit itself, which includes the net list, the device list,
   the model parameter list, etc.
2. Direct output from some of the analyses without the .PLOT and .PRINT
   commands. This includes the output from .OP, .TF, .SENS, .NOISE, and
   .FOUR analyses.
3. Print and plot by .PLOT and .PRINT commands. This includes the output
   from the .DC, .AC, and .TRAN analyses.
4. Run statistics. This includes the various kinds of summary information
   about the whole run including times required by various analyses and the
   amount of memory used.

Example 2.1

The circuit in Figure 2.1 is to be simulated on *PSpice* to calculate and plot the
transient response from 0 to 3 ms with a time increment of 10 μs. The capacitor
voltage is the output, $V(3, 0) = V(3)$. $V(3)$ and $V(7)$ are to be plotted. The circuit file
is to be stored in file EX2-1.CIR and the outputs are to be stored in file EX2-1.OUT.
The results should also be available for display and hard copy using the .PROBE
command.

Solution   The circuit file contains the following statements:
Example 2.1   A RLC circuit with sinusoidal input voltage
* The format for simple sinusoidal source is
* SIN   (VO   VA   FREQ)
* Refer to Chapter 4 for modeling sources
* VIN is connected between nodes 7 and 0, assuming that node 7 is at a higher
* potential with respect to node 0
* With peak voltage of VA = 10 V, frequency of FREQ = 1 kHz and offset value of
* VO = 0, the source is described by
    VIN   7   0   SIN (0   10   1KHZ)
* R1 with a value of 2 Ω is connected between nodes 7 and 5.
* Assuming that current flows into R1 from node 7 to node 5 and the voltage of node
* 7 with respect to node 5, V(7, 5) is positive, R1 is described by
    R1   7   5   2
* L1 with a value of 1.5 mH is connected between nodes 5 and 3.
* Assuming that current flows into L1 from node 5 to node 3 and the voltage of node
* 5 with respect to node 3, V(5, 3) is positive, L1 is described by,
    L1   5   3   1.5MH
* C1 with a value of 10 μF is connected between nodes 3 and 0.
* Assuming that current flows into C1 from node 3 to node 0 and the voltage of node
* 3 with respect to node 0, V(3) is positive, C1 is described by
    C1   3   0   10UF
* Transient analysis is invoked by .TRAN command, whose simple format is
* .TRAN TSTEP TSTOP
* Refer to Chapter 6 for dot commands. For transient analysis from 0 to 3 ms with

* an increment of 10 μs, the statement is
  ```
  .TRAN  10US  3MS
  ```
* Plots the results of transient analysis — V(3) and V(7)
  ```
  .PLOT  TRAN  V(3)  V(7)
  ```
* Graphic output can be obtained by simply invoking .PROBE command.
* Refer to Chapter 6 for dot commands.
  ```
  .PROBE
  ```
* The end of program is invoked by .END command
  ```
  .END
  ```

If the *PSpice* programs are loaded in a fixed disk and the circuit file is stored in a floppy diskette on drive A, the general command to run the circuit file is

```
PSPICE a:⟨input file⟩  a:⟨output file⟩
```

For an input file EX2-1.CIR and the output file EX2-1.OUT, the command is

```
PSPICE a:EX2-1.CIR  a:EX2-1.OUT
```

If the output file name is omitted, the results are stored by default on an output file that has the same name as the input file and is in the same drive but with an extension of .OUT.

It is a good practice to have .CIR and .OUT extensions on files so that the circuit file and the corresponding output file can be identified. Thus, the command can simply be

```
PSPICE a:EX2-1.CIR
```

The results of the transient response that are obtained on the display by .PROBE command are shown in Figure 2.2. The results of the .PRINT statement can be obtained by printing the contents of the output file EX2-1.OUT.

Example 2.2

Repeat Example 2.1 if the input is a step voltage as shown in Figure 2.3.

Solution   The circuit file is similar to that in Example 2.1, except the input is a step signal instead of sine wave. The step signal can be represented by a piecewise linear source and it is described, in general, by

```
PWL (T1  V1  T2  V2 ....... TN  VN)
```

where VN is the voltage at time TN.

Assuming a rise of 1 ns, the step voltage in Figure 2.3 can be described by

```
PWL (0  0  1NS  1V  4MS  1V)
```

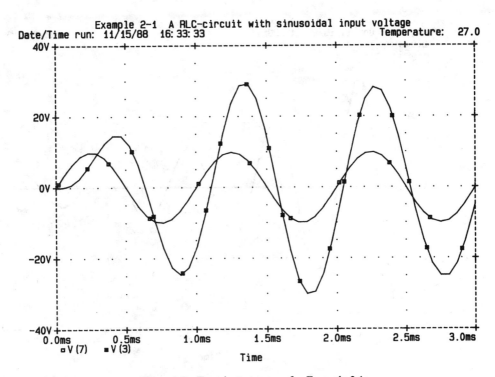

Figure 2.2   Transient response for Example 2.1

The circuit file contains the following statements:

Example 2.2   Step response of RLC circuit
* Refer to Section 4.2.3 for modeling piecewise linear source.
```
VIN   7   0    PWL (0   0   1NS   1V   4MS   1V )
R1    7   5    2
L1    5   3    1.5MH
C1    3   0    10UF
.TRAN   10US   4MS
.PRINT   TRAN   V(3)   V(7)
.PROBE
.END
```

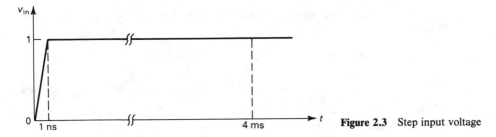

Figure 2.3   Step input voltage

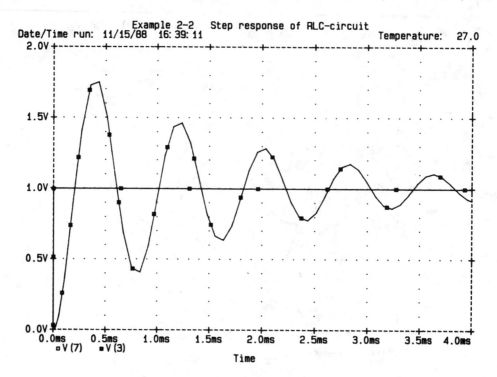

Figure 2.4   Step response of RLC circuit for Example 2.2

The results of the transient response that are obtained on the display by .PROBE command are shown in Figure 2.4. The results of the .PRINT statement can be obtained by printing the contents of the output file EX2-2.OUT, from the "Browse Menu."

Example 2.3

For the circuit in Fig. 2.1, the frequency response is to be calculated and printed over the frequency range from 10 Hz to 100 kHz with a decade increment and 10 points per decade. The peak magnitude and phase angle of the voltage across the capacitor are to be plotted on the output file. The results should also be available for display and hard copy using the .PROBE command.

Solution   The circuit file is similar to that in Example 2.1, except the statements for analysis and output are different.

The frequency response analysis is invoked by the .AC command, whose format is

```
.AC    DEC   NP   FSTART   FSTOP
```

where

DEC   = sweep by decade
NP   = number of points per decade

FSTART = starting frequency
FSTOP = ending (or stop) frequency

For NP = 10, FSTART = 10 Hz, and FSTOP = 100 kHz, the statement is

```
.AC  DEC  10  10  100KHZ
```

The magnitude and phase of voltage V(3) are specified as VM(3) and VP(3). The statement to plot is

```
.PLOT  AC  VM(3)  VP(3)        ·
```

The input voltage is ac type and the frequency is variable. We can consider a voltage source with a peak magnitude of 1V. The statement for an independent voltage source is

```
VIN 7  0  AC  1V
```

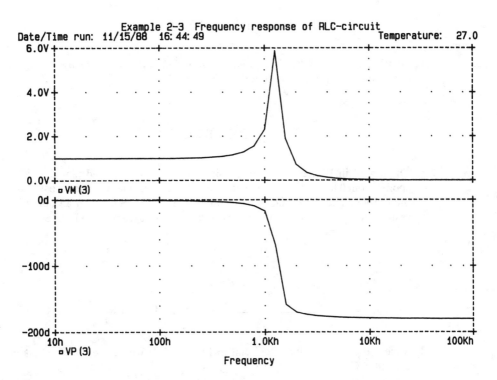

Figure 2.5   Frequency responses of an RLC circuit for Example 2.3

The circuit file contains the following statements:

Example 2.3   Frequency response of RLC circuit

* VIN is an independent voltage source whose frequency is varied by *PSpice* during
* the frequency response analysis.

```
VIN   7   0   AC   1V
R1    7   5   2
L1    5   3   1.5MH
C1    3   0   10UF
```

* The frequency response analysis is invoked by .AC command, whose format is
* .AC   DEC   NP   FSTART   FSTOP
* Refer to Chapter 6 for dot commands.

```
.AC   DEC   10   10   100KHZ
```

* Plot the results of .AC analysis − magnitude and phase of V(3).

```
.PLOT   AC   VM(3)   VP(3)
.PROBE
.END
```

The results of the frequency response that are obtained on the display by .PROBE command are shown in Figure 2.5. The results of the .PLOT statement can be obtained by printing the contents of the output file EX2-3.OUT.

REFERENCE

1. Paul W. Tuinenga, *SPICE: A guide to circuit simulation and analysis using PSpice.* Englewood Cliffs, N.J.: Prentice Hall Inc., 1988.

PROBLEMS

2.1. The circuit in Figure P2.1 is to be simulated to calculate and plot the transient response from 0 to 3 ms with a time increment of 10 μs. The voltage across resistor R_1 is the output. The input and output voltages are to be plotted on an output file. The results should also be available for display and hard copy by .PROBE command.

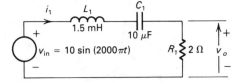

Figure P2.1

2.2. Repeat Problem 2.1 for the circuit in Figure P2.2, where the output is taken across capacitor C_1.

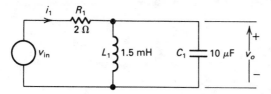

Figure P2.2

2.3. Repeat Problem 2.1 if the input is a step input, as shown in Figure 2.3.

2.4. Repeat Problem 2.2 if the input is a step input, as shown in Figure 2.3.

2.5. The circuit in Figure P2.1 is to be simulated to calculate and print the frequency response over the frequency range from 10 Hz to 100 kHz with a decade increment and 10 points per decade. The peak magnitude and phase angle of the voltage across the resistor is to be printed on the output file. The results should also be available for display and hard copy using the .PROBE command.

2.6. Repeat Problem 2.5 for the circuit in Figure P2.2, where the output is taken across capacitor C_1.

3

Defining Output Variables

3.1 INTRODUCTION

PSpice has some unique features of printing or plotting output voltages or currents by .PRINT and .PLOT statements. The .PRINT and .PLOT statements, which may have up to 8 output variables, are discussed in Chapter 6. The output variables that are allowed in .PRINT and .PLOT statements depend on the types of analyses:

> DC sweep and transient analysis
> AC analysis
> Noise analysis

3.2 DC SWEEP AND TRANSIENT ANALYSIS

DC sweep and transient analysis use similar type of output variables. The output variables can be divided into two types: voltage output and current output. An output variable can be assigned the symbol of a device (or element) or the terminal symbol of a device to identify whether the output is the voltage across the device

TABLE 3.1   SYMBOLS OF TWO-TERMINAL ELEMENTS

First letter	Element
C	Capacitor
D	Diode
E	Voltage-controlled voltage source
F	Current-controlled current source
G	Voltage-controlled current source
H	Current-controlled voltage source
I	Independent current source
L	Inductor
R	Resistor
V	Independent voltage source

TABLE 3.2   SYMBOLS AND TERMINAL SYMBOLS OF THREE- OR FOUR-TERMINAL DEVICES

First letter	Device	Terminals
B	GaAs MESFET	D (Drain)
		G (Gate)
		S (Source)
J	JFET	D (Drain)
		G (Gate)
		S (Source)
M	MOSFET	D (Drain)
		G (Gate)
		S (Source)
		B (Bulk, substrate)
Q	BJT	C (Collector)
		B (Base)
		E (Emitter)
		S (Substrate)

or current through the device (or element). Table 3.1 shows the symbols of two-terminal elements. Table 3.2 shows the symbols and terminal symbols of three- or four-terminal devices.

3.2.1 Voltage Output

For DC sweep and transient analysis, the output voltages can be obtained by the following statements:

V(⟨node⟩)	Voltage at ⟨node⟩ with respect to ground
V(N1, N2)	Voltage at node N_1 with respect to node N_2
V(⟨name⟩)	Voltage across two-terminal device, ⟨name⟩

Vx(⟨name⟩)	Voltage at terminal x of three-terminal device, ⟨name⟩
Vxy(⟨name⟩)	Voltage across terminals x and y of three-terminal device, ⟨name⟩
Vz(⟨name⟩)	Voltage at port z of transmission line, ⟨name⟩

VARIABLES	MEANING
V(5)	Voltage at node 5 with respect to ground.
V(4,2)	Voltage of node 4 with respect to node 2.
V(R1)	Voltage of resistor R_1, where the first node (as defined in the circuit file) is positive with respect to the second node.
V(D1)	Voltage across diode D_1 where the anode is positive with respect to cathode.
VC(Q3)	Voltage at the collector of transistor Q_3 with respect to ground.
VDS(M6)	Drain-source voltage of MOSFET M_6.
VB(T1)	Voltage at port B of transmission line T_1.

Note. SPICE and some versions of *PSpice* do not permit measuring voltage across a resistor, e.g., V(R1).

3.2.2 Current Output

For DC sweep and transient analysis, the output currents can be obtained by the following statements:

I(⟨name⟩)	Current through ⟨name⟩
Ix(⟨name⟩)	Current into terminal x of ⟨name⟩
Iz(⟨name⟩)	Current at port z of transmission line, ⟨name⟩

VARIABLES	MEANING
I(VS)	Current flowing into dc source, V_S.
I(R5)	Current flowing into resistor R_5 where the current is assumed to flow from the first node (as defined in the circuit file) through R_5 to the second node.
I(D1)	Current into diode D_1.
IC(Q4)	Current into the collector of transistor Q_4.
IG(J1)	Current into gate of JFET J_1.
ID(M5)	Current into drain of MOSFET M_5.
IA(T1)	Current at port A of transmission line T_1.

Note. SPICE and some versions of *PSpice* do not permit measuring the current through a resistor, e.g., *I*(*R5*). The easiest way is to add a dummy voltage source of 0 V (say, $V_X = 0$ V) and to measure the current through that source, e.g., *I*(*VX*).

Example 3.1

For the circuit in Figure 3.1, write the various currents and voltages in forms that are allowed by *PSpice*. The DC sources of 0 V are introduced to measure currents I_1 and I_2.

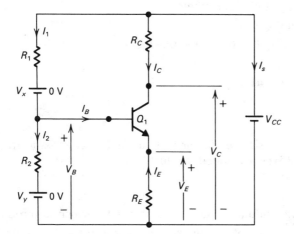

Figure 3.1 Bipolar transistor circuit

Solution

***PSpice* VARIABLES**

I_B	IB(Q1)	The base current of transistor Q_1
I_C	IC(Q1)	The collector current of transistor Q_1
I_E	IE(Q1)	The emitter current of transistor Q_1
I_S	I(VCC)	The current through voltage source V_{CC}
I_1	I(VX)	The current through voltage source V_X
I_2	I(VY)	The current through voltage source V_Y
V_B	VB(Q1)	The voltage at the base of transistor Q_1
V_C	VC(Q1)	The voltage at the collector of transistor Q_1
V_E	VE(Q1)	The voltage at the emitter of transistor Q_1
V_{CE}	VCE(Q1)	The collector-emitter voltage of transistor Q_1
V_{BE}	VBE(Q1)	The base-emitter voltage of transistor Q_1

3.3 AC ANALYSIS

In AC analysis, the output variables are sinusoidal quantities and are represented by complex numbers. An output variable can have magnitude, magnitude in decibels, phase, group delay, real part, and imaginary part. The output variables listed in Sections 3.2.1 and 3.2.2 are augmented by adding a suffix as follows:

SUFFIX	MEANING
(none)	Peak magnitude
M	Peak magnitude
DB	Peak magnitude in decibels
P	Phase in degrees
G	Group delay (ΔPHASE/ΔFREQUENCY)
R	Real part
I	Imaginary part

3.3.1 Voltage Output

The statements for AC analysis are similar to that for DC sweep and transient analysis, provided the suffixes are added as illustrated as follows:

VARIABLES

VM(5)	Magnitude of voltage at node 5 with respect to ground.
VM(4,2)	Magnitude of voltage at node 4 with respect to node 2.
VDB(R1)	DB magnitude of voltage across resistor R_1, where the first node (as defined in the circuit file) is assumed to be positive with respect to second node.
VP(D1)	Phase of anode voltage of diode D_1 with respect to cathode.
VCM(Q3)	Magnitude of the collector voltage of transistor Q_3 with respect to ground.
VDSP(M6)	Phase of the drain-source voltage of MOSFET' M_6.
VBP(T1)	Phase of voltage at port B of transmission line T_1.
VR(2,3)	Real part of voltage at node 2 with respect to node 3.
VI(2,3)	Imaginary part of voltage at node 2 with respect to node 3.

3.3.2 Current Output

The statements for AC analysis are similar to that for DC sweep and transient responses. However, only the currents through the elements in Table 3.3 are available.

TABLE 3.3 CURRENT THROUGH
ELEMENTS FOR AC ANALYSIS

First letter	Element
C	Capacitor
I	Independent current source
L	Inductor
R	Resistor
T	Transmission line
V	Independent voltage source

For all other elements, a zero-valued voltage source must be placed in series with the device (or device terminal) of interest. Then a print or plot statement should be used for the current through this voltage source.

VARIABLES	MEANING
IM(R5)	Magnitude of current through resistor R_5.
IR(R5)	Real part of current through resistor R_5.
II(R5)	Imaginary part of current through resistor R_5.
IM(VIN)	Magnitude of current through source v_{IN}.
IR(VIN)	Real part of current through source v_{IN}.
II(VIN)	Imaginary part of current through source v_{IN}.
IAG(T1)	Group delay of current at port A of transmission line T_1.

Example 3.2

For the circuit in Figure 3.2, write the various voltages and currents in forms that are allowed by *PSpice*. The dummy voltage source of 0 V is introduced to measure current, I_L.

Solution

PSpice VARIABLES		MEANING
V_2	VM(2)	The peak magnitude of voltage at node 2
$\underline{/V_2}$	VP(2)	The phase angle of voltage at node 2
V_{12}	VM(1,2)	The peak magnitude of voltage between nodes 1 and 2
$\underline{/V_{12}}$	VP(1,2)	The phase angle of voltage between nodes 1 and 2
I_R	IM(VX)	The magnitude of current through voltage source, v_X
$\underline{/I_R}$	IP(VX)	The phase angle of current through voltage source, v_X

I_L	IM(L1)	The magnitude of current through inductor L_1
$\underline{/I_L}$	IP(L1)	The phase angle of current through inductor L_1
I_C	IM(C1)	The magnitude of current through capacitor C_1
$\underline{/I_C}$	IP(C1)	The phase angle of current through capacitor C_1

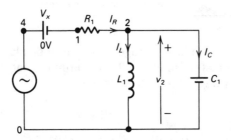

Figure 3.2 Circuit for Example 3.2

3.4 NOISE ANALYSIS

For the noise analysis, the output variables are predefined as follows:

OUTPUT VARIABLES	MEANING
ONOISE	Total RMS summed noise at output node
INOISE	ONOISE equivalent at the input node
DB(ONOISE)	ONOISE in decibels
DB(INOISE)	INOISE in decibels

Noise Output Statement

```
.PRINT NOISE INOISE ONOISE
```

Note. The noise output from only one device can not be obtained by .PRINT or .PLOT command. However, the print interval on the .NOISE statement can be used to output this information. The .NOISE command is discussed in Section 6.8.

SUMMARY

V(⟨node⟩)	Voltage at ⟨node⟩ with respect to ground
V(N1, N2)	Voltage at node N_1 with respect to node N_2
V(⟨name⟩)	Voltage across two-terminal device ⟨name⟩

Vx(⟨name⟩)	Voltage at terminal x of device ⟨name⟩
Vxy(⟨name⟩)	Voltage at terminal x with respect to terminal y for device ⟨name⟩
Vz(⟨name⟩)	Voltage at port z of transmission line ⟨name⟩
I(⟨name⟩)	Current through device ⟨name⟩
Ix(⟨name⟩)	Current into terminal x of device ⟨name⟩
Iz(⟨name⟩)	Current at port z of transmission line ⟨name⟩
(none)	Magnitude
M	Magnitude
DB	Magnitude in decibels
P	Phase in degrees
G	Group delay (Δphase/Δfrequency)
R	Real part
I	Imaginary part

— 4 —

Voltage and Current Sources

4.1 INTRODUCTION

PSpice allows generation of dependent (or independent) voltage and current sources. Independent source can be time variant. A nonlinear source can also be simulated by a polynomial. This chapter explains the techniques for generating simulating sources. The *PSpice* statements for various sources require:

> Source modeling
> Independent sources
> Dependent sources

4.2 SOURCES MODELING

The independent voltage and current sources that can be modeled by *PSpice* are

> Exponential
> Pulse
> Sinusoidal

Piecewise linear

Single-frequency frequency-modulation

Polynomial

4.2.1 Exponential Source

The waveform and parameters of an exponential waveform are shown in Figure 4.1 and Table 4.1. The symbol of exponential sources is EXP and the general form is

```
EXP (V1   V2   TRD   TRC   TFD   TFC)
```

V_1 and V_2 *must* be specified by the user. (TSTEP is the incrementing time during transient (.TRAN) analysis.) In an EXP waveform, the voltage remains V_1 for the first TRD seconds. Then the voltage rises exponentially from V_1 to V_2 with a rise time constant of TRC. After a time of TFD, the voltage falls exponentially from V_2 to V_1 with a fall time constant of TFC.

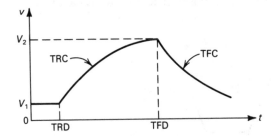

Figure 4.1 Exponential waveform

Note. The values of EXP waveform as well as the values of other time-dependent waveforms at intermediate time points are determined by *PSpice* by means of linear interpolation.

Typical Model Statements.   For $V_1 = 0$, $V_2 = 1$ V, TRD = 2 ns, TRC = 20 ns, TFD = 60 ns, and TFC = 30 ns, the model statement is

```
EXP (0   1   2NS   20NS   60NS   30NS)
```

TABLE 4.1 MODEL PARAMETERS OF EXP SOURCES

Name	Meaning	Units	Default
V1	Initial voltage	Volts	None
V2	Pulsed voltage	Volts	None
TRD	Rise delay time	Seconds	0
TRC	Rise time constant	Seconds	TSTEP
TFD	Fall delay time	Seconds	TRD + TSTEP
TFC	Fall time constant	Seconds	TSTEP

With TRD = 0, the statement becomes

```
EXP  (0   1   0     20NS   60NS   30NS)
```

With $V_1 = -1$ V and $V_2 = 2$ V, it is

```
EXP  (-1   2  2NS   20NS   60NS   30NS)
```

4.2.2 Pulse Source

The waveform and parameters of a pulse waveform are shown in Figure 4.2 and Table 4.2. The symbol of a pulse source is PULSE and the general form is

```
PULSE (V1   V2   TD   TR   TF   PW   PER)
```

V_1 and V_2 *must* be specified by the user. TSTEP and TSTOP are the incrementing time and stop time, respectively, during transient (.TRAN) analysis.

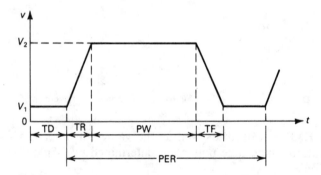

Figure 4.2   Pulse waveform

TABLE 4.2   MODEL PARAMETERS OF PULSE SOURCES

Name	Meaning	Units	Default
V1	Initial voltage	Volts	None
V2	Pulsed voltage	Volts	None
TD	Delay time	Seconds	0
TR	Rise time	Seconds	TSTEP
TF	Fall time	Seconds	TSTEP
PW	Pulse width	Seconds	TSTOP
PER	Period	Seconds	TSTOP

Typical Statements.   For $V_1 = -1$ V, $V_2 = 1$ V, TD = 2 ns, TR = 2 ns, TF = 2 ns, PW = 50 ns, and PER = 100 ns, the model statement is

```
PULSE (-1  1  2NS  2NS  2NS  50NS  100NS)
```

With $V_1 = 0$, $V_2 = 1$V, the model becomes

```
PULSE (0   1  2NS  2NS  2NS  50NS  100NS)
```

With $V_1 = 0$, $V_2 = -1$V, the model becomes

```
PULSE (0  -1  2NS  2NS  2NS  50NS  100NS)
```

4.2.3 Piecewise Linear Source

A point in a waveform can be described by (T_i, V_i), and every pair of values (T_i, V_i) specifies the source value at time T_i. The voltage at times between the intermediate points is determined by *PSpice* by using linear interpolation. The symbol of a piecewise linear source is PWL and the general form is

```
PWL (T1  V1  T2  V2 ....... TN  VN)
```

The model parameters of PWL waveform are given in Table 4.3.

TABLE 4.3   MODEL PARAMETERS OF PWL SOURCES

Name	Meaning	Units	Default
Ti	Time at a point	Seconds	None
Vi	Voltage at a point	Seconds	None

Typical Statement.   The model statement for the typical waveform in Figure 4.3 is

```
PWL (0  3  10US  3V  15US  6V  40US  6V  45US  2V  60US  2V)
```

4.2.4 Single-Frequency Frequency-Modulation

The symbol of a source with single-frequency frequency-modulation is SFFM, and the general form is

```
SFFM (VO  VA  FC  MOD  FS)
```

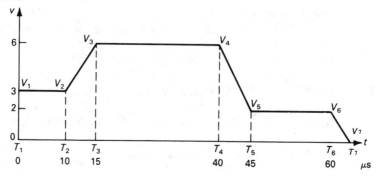

Figure 4.3   Piecewise linear waveform

The model parameters of SFFM waveform are given in Table 4.4

TABLE 4.4   Model Parameters of SFFM Sources

Name	Meaning	Units	Default
VO	Offset voltage	Volts	None
VA	Amplitude of voltage	Volts	None
FC	Carrier frequency	Hertz	1/TSTOP
MOD	Modulation index		0
FS	Signal frequency	Hertz	1/TSTOP

V_O and V_A *must* be specified by the user. TSTOP is the stop time during transient (.TRAN) analysis. The waveform is of the form

$$V = V_O + V_A \sin[(2\pi F_C t) + M \sin(2\pi F_S t)]$$

Typical Statements.   For $V_O = 0$, $V_A = 1\text{V}$, $F_C = 30$ MHz, MOD = 5, and $F_S = 5$ *kHz*, the model statement is

```
SFFM (0    1V    30MHZ    5    5KHZ)
```

With $V_O = 1$ mV and $V_A = 2$ V, the model becomes

```
SFFM (1MV   2V   30MHZ   5   5KHZ)
```

4.2.5 Sinusoidal Source

The symbol of a sinusoidal source is SIN, and the general form is

```
SIN (VO  VA  FREQ  TD  ALP  THETA)
```

The model parameters of SIN waveform are given in Table 4.5.

TABLE 4.5   MODEL PARAMETERS OF SIN SOURCES

Name	Meaning	Units	Default
VO	Offset voltage	Volts	None
VA	Peak voltage	Volts	None
FREQ	Frequency	Hertz	1/TSTOP
TD	Delay time	Seconds	0
ALPHA	Damping factor	1/seconds	0
THETA	Phase delay	Degrees	0

V_O and V_A *must* be specified by the user. TSTOP is the stop time during transient (.TRAN) analysis. The waveform stays at 0 for a time of TD and then the voltage becomes an exponentially damped sine wave. An exponentially damped sine wave is described by

$$V = V_O + V_A e^{-\alpha(t - t_d)} \sin[(2\pi f(t - t_d) - \theta]$$

and this is shown in Figure 4.4.

Typical Statements

```
SIN (0   1V   10KHZ   10US   1E5 )
SIN (1   5V   10KHZ   1E5    30DEG )
SIN (0   2V   10KHZ   30DEG )
SIN (0   2V   10KHZ )
```

4.2.6 Polynomial Source

The symbol of a polynomial or nonlinear source is POLY(*n*), where *n* is the number of dimensions of the polynomial. The default value of *n* is 1. The dimen-

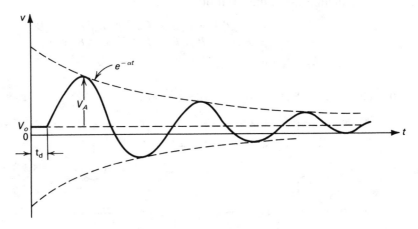

Figure 4.4   Damped sinusoidal waveform

sions depend on the number of controlling sources. The general form is

```
POLY (n) ⟨(controlling) nodes⟩ ⟨(coefficients) values⟩
```

The output sources or the controlling sources can be voltages or currents. For voltage-controlled sources, the number of controlling nodes must be twice the number of dimensions. For current-controlled sources, the number of controlling sources must be equal to the number of dimensions. The number of dimensions and the number of coefficients are arbitrary.

Let us call A, B, and C the three controlling variables and Y the output source. Figure 4.5 shows a source that is controlled by A, B, and C. The output source takes the form of

$$Y = f(A, B, C, \ldots)$$

where

Y can be voltage or current

A, B, and C can be voltage or current or any combination

For a polynomial of $n = 1$ with A as the only controlling variable, the source function takes the form of

$$Y = P_0 + P_1 A + P_2 A^2 + P_3 A^3 + P_4 A^4 + \cdots + P_n A^n$$

where $P_0, P_1, \ldots, P_n$ are the coefficients values; this is written in *PSpice* as

```
POLY   NC1+    NC1-    P0   P1   P2   P3   P4   P5 ... Pn
```

where NC1+ and NC1− are the positive and negative nodes, respectively, of controlling source A.

For a polynomial of degree $n = 2$ with A and B as the controlling sources, the source function takes the form of

$$Y = P_0 + P_1 A + P_2 B + P_3 A^2 + P_4 AB + P_5 B^2$$
$$+ P_6 A^3 + P_7 A^2 B + P_8 AB^2 + P_9 B^3 + \cdots$$

This is described in *PSpice* as

```
POLY(2) NC1+   NC1-   NC2+   NC2-   P0   P1   P2   P3   P4   P5 ... Pn
```

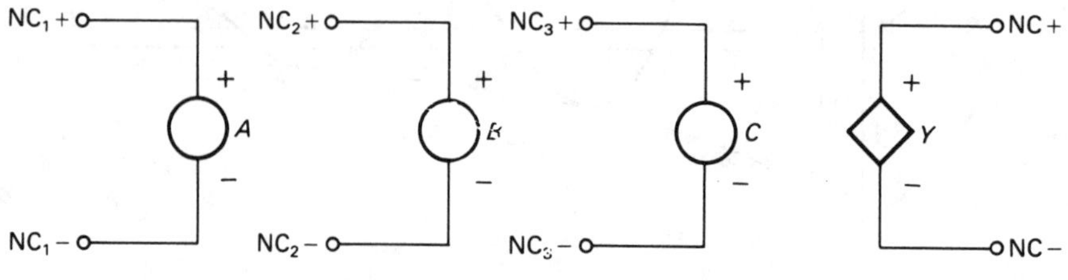

Figure 4.5 Polynomial source

where NC1+, NC2+ and NC1−, NC2− are the positive and negative nodes, respectively, of the controlling sources.

For a polynomial of degree $n = 3$ with A, B, and C as the controlling sources, the source function takes the form of

$$Y = P_0 + P_1A + P_2B + P_3C + P_4A^2 + P_5AB + P_6AC$$
$$+ P_7B^2 + P_8BC + P_9C^2 + P_{10}A^3 + P_{11}A^2B$$
$$+ P_{12}A^2C + P_{13}AB^2 + P_{14}ABC + P_{15}AC^2$$
$$+ P_{16}B^3 + P_{17}B^2C + P_{18}BC^2 + P_{19}C^3 + P_{20}A^4 + \cdots$$

This is written in *PSpice* as

```
POLY(3) NC1+ NC1- NC2+ NC2- NC3+ NC3- P₀  P₁  P₂  P₃  P₄  P₅ ... Pₙ
```

where NC1+ NC2+, NC3+ and NC1−, NC2−, NC3− are the positive and negative nodes, respectively, of the controlling sources.

   Typical Model Statements.   For $Y = 2\,V(10)$, the model is

```
POLY    10    0    2.0
```

For $Y = V(5) + [V(5)]^2 + [V(5)]^3 + [V(5)]^4$, the model is

```
POLY      5    0    0.0   1.0   1.0   1.0   1.0
```

For $Y = V(3) + V(5) + [V(3)]^2 + V(3)\,V(5)$, the model is

```
POLY(2)   3    0    5    0   0.0   1.0   1.0   1.0   1.0
```

For $Y = V(3) + V(5) + V(10) + [V(3)]^2$, the model is

```
POLY(3)   3    0    5    0   10   0   0.0   1.0   1.0   1.0   1.0
```

If $I(VN)$ is the controlling current through voltage source V_N and $Y = I(VN) + [I(VN)]^2 + [I(VN)]^3 + [I(VN)]^4$, the model is

```
POLY    VN    0.0   1.0   1.0   1.0   1.0
```

If $I(VN)$ and $I(VX)$ are the controlling currents and $Y = I(VN) + I(VX) + [I(VN)]^2 + I(VN)\,I(VX)$.

```
POLY(2)   VN   VX   0.0   1.0   1.0   1.0   1.0
```

   Note. If the source is of one dimension and only one coefficient is specified as in the first example, *PSpice* assumes $P_0 = 0$ and the specified value as P_1. That is, $Y = 2A$.

4.3 INDEPENDENT SOURCES

The independent sources can be time invariant and time variant. They can be currents or voltages, as shown in Figure 4.6.

4.3.1 Independent Voltage Source

The symbol of an independent voltage source is V and the general form is

```
V⟨name⟩ N+  N-  [DC ⟨value⟩]  [AC ⟨(magnitude) value⟩ ⟨(phase) value⟩]
+      [(transient value)
+      [PULSE] [SIN] [EXP] [PWL] [SFFM] [source arguments]]
```

N+ is the positive node and N− is the negative node, as shown in Figure 4.6(a). Positive current flows from node N+ through the voltage source to the negative node N−. The voltage source need not be grounded. For the DC, AC, and transient values, the default value is zero. None or all DC, AC, and transient values may be specified. The ⟨(phase) value⟩ is in degrees.

The source is set to the *DC* value in *DC* analysis. It is set to *AC* value in *AC* analysis. If the ⟨(phase) value⟩ in *AC* analysis is omitted, the default is 0. The time-dependent source (e.g., PULSE, EXP, or SIN) is assigned for transient analysis. A voltage source may be used as an **ammeter** in *PSpice* by inserting a zero-valued voltage source into the circuit for the purpose of measuring current. Since a zero-valued source behaves as short circuit, there will be no effect on the circuit operation.

Typical Statements

```
V1      15   0   6V
V2      15   0   DC  6V
VAC     5    6   AC  1V
VACP    5    6   AC  1V   45DEG
VPULSE  10   0   PULSE (0  1  2NS  2NS  2NS  50NS  100NS)
VIN     25   22  DC  2 AC  1   30  SIN (0  2V  10KHZ)
```

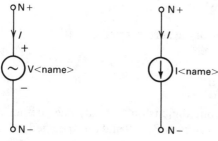

(a) Voltage source   (b) Current source   **Figure 4.6** Voltage and current sources

4.3.2 Independent Current Source

The symbol of an independent current source is *I* and the general form is

```
I(name) N+  N-  [DC (value)]  [AC ((magnitude) value) ((phase) value)]
+       [(transient value)
+       [PULSE] [SIN] [EXP] [PWL] [SFFM] [source arguments]]
```

Note. The first column with + *(plus)* signifies continuation of the *PSpice* statement. After the + sign, the statement can continue in any column.

N+ is the positive node and N− is the negative node, as shown in Figure 4.6(b). Positive current flows from node N+ through the current source to the negative node N−. The current source need not be grounded. The source specifications are similar to that of independent voltage source.

Typical Statements

```
I1      15   0   2.5MA
I2      15   0   DC  2.5MA
IAC     5    6   AC  1V
IACP    5    6   AC  1V   45DEG
IPULSE  10   0   PULSE (0   1V   2NS   2NS   2NS   50NS   100NS)
IIN     25   22  DC  2   AC  1V   30DEG   SIN (0   2V   10KHZ)
```

4.4 DEPENDENT SOURCES

The four types of dependent sources are shown in Figure 4.7. They are

> Voltage-controlled voltage source
> Voltage-controlled current source
> Current-controlled current source
> Current-controlled voltage source

4.4.1 Voltage-Controlled Voltage Source

The symbol of a voltage-controlled voltage source in Figure 4.7(a) is *E*; it takes a linear form as in

```
E(name)   N+ N-   NC+   NC-   ((voltage gain) value)
```

N+ and N− are the positive and negative output nodes, respectively. NC+ and NC− are the positive and negative nodes, respectively, of the controlling voltage.

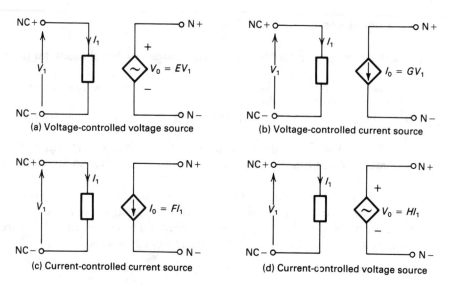

Figure 4.7 Dependent sources

The nonlinear form is

```
E<name>   N+ N-   [POLY(<value>)]
+             <<(+ controlling) node><(- controlling) node>> (pairs)
+             <(polynomial coefficients) values>]
```

The POLY source is described in Section 4.2.6. The number of controlling nodes
is twice the number of dimensions. A particular node may appear more than once
and the output and the controlling nodes could be the same.

Typical Statements

```
EAB       1   2   4   6   1.0
EVOLT     4   7   20  22  2E5
ENONLIN   25  40  POLY(2) 3  0  5  0  0.0  1.0  1.5  1.2  1.7
E2        10  12  POLY  5  0  0.0  1.0  1.5  1.2  1.7
```

Notes

1. The source ENONLIN that specifies a polynomial voltage source between
   nodes 25 and 40 is controlled by V(3), and V(5). Its value is given by

$$Y = V(3) + 1.5V(5) + 1.2[V(3)]^2 + 1.7V(3)V(5)$$

The source E2 that specifies a polynomial voltage source between nodes 10
and 12 is controlled by V(5, 0) and is given by

$$Y = V(5) + 1.5[V(5)]^2 + 1.2[V(5)]^3 + 1.7[V(5)]^4$$

4.4.2 Voltage-Controlled Current Source

The symbol of a voltage-controlled current source as shown in Figure 4.7(b) is G; its linear form is

```
G⟨name⟩   N+ N-   NC+   NC-   ⟨(transconductance) value⟩
```

N+ and N- are the positive and negative output nodes, respectively. NC+ and NC- are the positive and negative nodes, respectively, of the controlling voltage. The nonlinear form is

```
G⟨name⟩  N+ N-  [POLY(⟨value⟩)
+        ⟨⟨(+ controlling) node⟩ ⟨(- controlling) node⟩⟩ (pairs)
+        ⟨(polynomial coefficients) values⟩]
```

Typical Statements

```
GAB       1   2   4   6   1.0
GVOLT     4   7   20  22  2E5
GNONLIN   25  40 POLY(2) 3  0  5  0  0.0  1.0  1.5  1.2  1.7
G2        10  12 POLY  5  0  0.0  1.0  1.5  1.2  1.7
```

Notes

1. The source GNONLIN that specifies a polynomial current source from node 25 to node 40 is controlled by V(3), and V(5) and it is given by

$$I = V(3) + 1.5V(5) + 1.2[V(3)]^2 + 1.7V(3)V(5)$$

   The source G2 that specifies a polynomial current source from node 10 to node 12 is controlled by V(5); it is given by

$$I = V(5) + 1.5[V(5)]^2 + 1.2[V(5)]^3 + 1.7[V(5)]^4$$

2. A nonlinear conductance can be simulated by a voltage-controlled current source. A linear voltage-controlled current source is the same as a conductance if the controlling nodes are the same as the output nodes.

```
GRES   4   6   4   6   0.1
```

   is a conductance of 0.1 mhos with a resistance of 1/0.1 = 10 Ω.

```
GHMO   1   2   POLY   1   2   0.0 1.5M   1.7M
```

   represents

$$I = 1.5 \times 1^{-3}V(1,2) + 1.7 \times 10^{-3}[V(1,2)]^2$$

   and is a nonlinear conductance in mhos.

4.4.3 Current-Controlled Current Source

The symbol of a current-controlled current source, as shown in Figure 4.7(c), is F and its linear form is

```
F⟨name⟩   N+ N-   VN   ⟨(current gain) value⟩
```

N+ and N− are the positive and negative nodes, respectively, of the current source. *VN* is a voltage source through which the controlling current flows. The controlling current is assumed to flow from the positive node of *VN* through the voltage source *VN* to the negative node of *VN*. The current through the controlling voltage source I(VN) determines the output current. The voltage source *VN* that monitors the controlling current must be an independent voltage source and it can have a **zero** or finite value. The current through a resistor controls the source, a dummy voltage source of 0 V should be connected in series with the resistor to monitor the controlling the current.

The nonlinear form is

```
F⟨name⟩   N+ N-   [POLY(⟨value⟩
+               VN1, VN2, VN3.....
+               ⟨(polynomial coefficients) values⟩]
```

The POLY source is described in Section 4.2.6. The number of controlling current sources must be equal to the number of dimensions.

Typical Statements

```
FAB       1    2   VIN   10
FAMP      13   4   VCC   50
FNONLIN   25 40 POLY  VN   0.0   1.0   1.5   1.2   1.7
```

Note. The source FNONLIN that specifies a polynomial current source from node 25 to node 40 is given by

$$I = I(VN) + 1.5[I(VN)]^2 + 1.2[I(VN)]^3 + 1.7[I(VN)]^4$$

4.4.4 Current-Controlled Voltage Source

The symbol of a current-controlled voltage source, as shown in Figure 4.7(d), is H and its linear form is

```
H⟨name⟩   N+ N-   VN   ⟨(transresistance) value⟩
```

N+ and N− are the positive and negative nodes, respectively, of the voltage source. *VN* is a voltage source through which the controlling current flows, and its specifications are similar to that for current-controlled current source.

The nonlinear form is

```
H〈name〉  N+ N-  [POLY(〈value〉)
+        VN1, VN2, VN3 ...
+        〈(polynomial coefficients) values)]
```

Typical Statements

```
HAB      1    2    VIN   10
HAMP     13   4    VCC   50
HNONLIN  25   40   POLY  VN   0.0   1.0   1.5   1.2   1.7
```

Notes

1. The source HNONLIN that specifies a polynomial voltage source between nodes 25 and 40 is controlled by I(VN) and is given by

$$V = I(VN) + 1.5[I(VN)]^2 + 1.2[I(VN)]^3 + 1.7[I(VN)]^4$$

2. A nonlinear resistance can be simulated by a current-controlled voltage source. A linear current-controlled voltage source is the same as a resistor if the controlling current is the same as the current through the voltage between output nodes.

```
HRES   4   6   VN   10
```

is a resistance of 10 Ω.

```
HMHO   1   2   POLY   I(VN)   0.0   1.5M   1.7M
```

represents

$$H = 1.5 \times 1^{-3}I(VN) + 1.7 \times 10^{-3}[I(VN)]^2$$

and is a nonlinear resistance in ohms.

SUMMARY

EXP	Exponential source
	`EXP (V1  V2  TRD TRC TFD TFC)`
POLY	Polynomial source
	`POLY (n) ((controlling) nodes) ((coefficients) values)`
PULSE	Pulse source
	`PULSE (V1  V2  TD TR TF PW PER)`
PWL	Piecewise linear source
	`PWL (T1 V1 T2 V2 ... TN VN)`
SFFM	Single-frequency frequency-modulation
	`SFFM (VO  VA  FC MOD FS)`

SIN   Sinusoidal source
                       SIN (VO VA FREQ   TD ALP THETA)
E   Voltage-controlled voltage source
                       E(name) N+ N-   NC+ NC-   ((voltage gain) value)
F   Current-controlled current source
                       F(name) N+ N-   VN   ((current gain) value)
G   Voltage-controlled current source
                       G(name) N+ N-   NC+ NC-   ((transconductance) value)
H   Current-controlled voltage source
                       H(name) N+ N-   VN   ((transresistance value)
I   Independent current source
                       I(name) N+   N-   [DC (value)]   [AC ((magnitude) value)
                       +   ((phase) value)] [(transient value)
                       +   [PULSE] [SIN] [EXP] [PWL] [SFFM] [source arguments]]
V   Independent voltage source
                       V(name) N+   N-   [DC (value)]   [AC ((magnitude) value)
                       +   ((phase) value)] [(transient value)
                       +   [PULSE] [SIN] [EXP] [PWL] [SFFM] [source arguments]]

PROBLEMS

Write *PSpice* statements for the following questions. Assume that the first node is the positive terminal and the second node is the negative terminal.

4.1. The various voltage or current waveforms that are connected between nodes 4 and 5 are shown in Figure P4.1.

4.2. A voltage source that is connected between nodes 4 and 5 is given by

$$v = 2 \sin\{2\pi\ 50{,}000t + 5 \sin(2\pi\ 1000t)\}$$

4.3. A voltage source that is connected between nodes 10 and 0 has a DC voltage of 12 V for DC analysis, peak voltage of 2 V with 60° phase shift for AC analysis, and a sinusoidal peak voltage of 0.1 V at 1 MHz for transient analysis.

4.4. A current source that is connected between nodes 5 and 0 has a DC voltage of 0.1A for DC analysis, peak current of 1 A with 60° phase shift for AC analysis, and a sinusoidal current of 0.1 A at 1 kHz for transient analysis.

4.5. A polynomial voltage source Y that is connected between nodes 1 and 2 is controlled by a voltage source V_1 connected between nodes 4 and 5. The source is given by

$$Y = 0.1V_1 + 0.2V_1^2 + 0.05V_1^3$$

4.6. A polynomial current source I that is connected between nodes 1 and 2 is controlled by a voltage source V_1 connected between nodes 4 and 5. The source is given by

$$Y = 0.1V_1 + 0.2V_1^2 + 0.05V_1^3$$

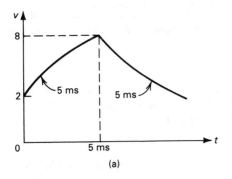

(a)

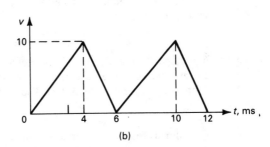

(b)

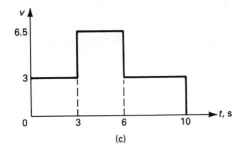

(c)

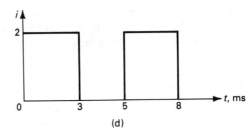

(d)

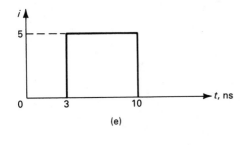

(e)

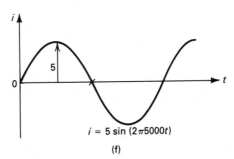

$i = 5 \sin (2\pi 5000t)$

(f)

Figure P4.1

4.7. A voltage source V_o that is connected between nodes 5 and 6 is controlled by a voltage source V_1 and has a voltage gain of 25. The controlling voltage is connected between nodes 10 and 12. The source is expressed as

$$V_o = 25V_1$$

4.8. A current source I_o that is connected between nodes 5 and 6 is controlled by a current source I_1 and has a current gain of 10. The voltage through which the controlling current flows is V_C. The current source is given by

$$I_o = 10I_1$$

4.9. A current source I_o that is connected between nodes 5 and 6 is controlled by a voltage source V_1 between nodes 8 and 9. The transconductance is 0.05 mhos. The current source is given by

$$I_o = 0.05V_i$$

4.10. A voltage source V_o that is connected between nodes 5 and 6 is controlled by a current source I_1 and has a transresistance of 150 Ω. The voltage through which the controlling current flows is V_C. The voltage source is expressed as

$$V_o = 150I_1$$

4.11. A nonlinear resistance R that is connected between nodes 4 and 6 is controlled by a voltage source V_1 and has a resistance of the form

$$R = V_1 + 0.2V_1^2$$

4.12. A nonlinear transconductance G_m that is connected between nodes 4 and 6 is controlled by a current source. The voltage through which the controlling current flows is V_1. The transconductance has the form

$$G_m = V_1 + 0.2V_1^2$$

5

Passive Elements

5.1 INTRODUCTION

PSpice recognizes passive elements by their symbols and their models. The elements can be resistor (R), inductor (L), capacitor (C), magnetic, or switches. The models are necessary to take into account the parameter variations, e.g., the value of a resistor depends on the operating temperature. The simulation of passive elements on *PSpice* requires specifying the following:

Modeling of elements
Operating temperature
RLC elements
Magnetic elements
Transmission lines
Switches

5.2 MODELING OF ELEMENTS

A model that specifies a set of parameters for an element is specified in *PSpice* by the .MODEL command. The same model can be used by one or more elements in the same circuit. The various .(*dot*) commands are explained in Chapter 6. The general form of the model statement is

```
.MODEL  MNAME  TYPE (P1=V1 P2=V2 P3=V2 ... PN=VN)
```

MNAME is the name of the model and must start with a letter. Although not necessary, it is advisable to make the first letter the symbol of the element (e.g., *R* for resistor, *L* for inductor). The list of symbols for elements is shown in Table 2.1.

P_1, P_2, . . . are the element parameters and V_1, V_2, . . . are their values, respectively. TYPE is the type name of the elements and must have the correct type as shown in Table 5.1. An element must have the correct model type name. That is, a resistor must have the type name of RES, but not the type IND or CAP. However, there can be more than one model of the same type in a circuit with different model names.

Some Model Statements

```
.MODEL  RLOAD   RES (R=1   TC1=0.02   TC2=0.005)
.MODEL  CPASS   CAP (C=1   VC1=0.01   VC2=0.002   TC1=0.02   TC2=0.005)
.MODEL  LFILTER IND (L=1   IL1=0.1    IL2=0.002   TC1=0.02   TC2=0.005)
.MODEL  DNOM    D   (IS=1E-6)
.MODEL  QOUT    NPN (BF=50  IS=1E-9)
```

5.3 OPERATING TEMPERATURE

The operating temperature of an analysis can be set to any desire value by the .TEMP command. The general form of the statement is

```
.TEMP  ((one or more temperature) values)
```

TABLE 5.1   TYPE NAME OF ELEMENTS

Type name	Element
RES	Resistor
CAP	Capacitor
D	Diode
IND	Inductor
NPN	NPN bipolar junction transistor
PNP	PNP bipolar junction transistor
NJF	N-channel junction FET
PJF	P-channel junction FET
NMOS	N-channel MOSFET
PMOS	P-channel MOSFET
GASFET	N-channel GaAs MOSFET
VSWITCH	Voltage-controlled switch
ISWITCH	Current-controlled switch
CORE	Nonlinear magnetic core (transformer)

The temperatures are in degrees Celsius. If more than one temperature is specified, then the analysis is performed for each temperature.

The model parameters are assumed to be measured at a nominal temperature, which, by default, is 27°C. The default nominal temperature of 27°C can be changed by the TNOM option in the .OPTIONS statements that are discussed in Section 6.5.

Some Temperature Statements

```
.TEMP  50
.TEMP  25  50
.TEMP  0   25  50  100
```

5.4 RLC ELEMENTS

The voltage and current relationships of resistor, inductor, and capacitor are shown in Figure 5.1.

5.4.1 Resistor

The symbol for a resistor is R. The name of a resistor must start with R and it takes the general form of

```
R〈name〉 N+  N-  RNAME  VALUE
```

A resistor does not have a polarity and the order of the nodes does not matter. However, by defining N+ as the positive node and N− as the negative node, the current is assumed to flow from node N+ through the resistor to node N−. RNAME is the model name that defines the parameters of the resistor. VALUE is the nominal value of the resistance.

Note. Some versions of *PSpice* or SPICE does not recognize the polarity of resistors and do not allow to refer to currents through resistors—e.g., $I(R_1)$.

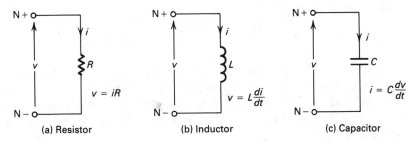

Figure 5.1 Voltage and current relationships

TABLE 5.2   MODEL PARAMETERS FOR RESISTORS

Name	Meaning	Units	Default
R	Resistance multiplier		1
TC1	Linear temperature coefficient	$°C^{-1}$	0
TC2	Quadratic temperature coefficient	$°C^{-2}$	0
TCE	Exponential temperature coefficient	$\% °C$	0

The model parameters are shown in Table 5.2. If RNAME is omitted, VALUE is the resistance in ohms and can be positive or negative but *must* not be zero. If RNAME is included and TCE is not specified, the resistance as a function of temperature is calculated from

$$RES = VALUE * R * [1 + TC1 * (T - T0) + TC2 * (T - T0)^2]$$

If RNAME is included and TCE is specified, the resistance as a function of temperature is calculated from

$$RES = VALUE * R * 1.01^{\,TCE\,*\,(T\,-\,T0)}$$

where T and T0 are the operating temperature and the room temperature, respectively, in degrees Celsius.

Some Resistor Statements

```
R1       6    5      10K
RLOAD    12   11     ARES   2MEG
.MODEL   ARES   RES (R=1   TC1=0.02   TC2=0.005)
RINPUT   15   14     BRES   5K
.MODEL   BRES   RES (R=1   TCE=2.5)
```

5.4.2 Capacitor

The symbol for a capacitor is C. The name of a capacitor must start with C and it takes the general form of

```
C(name)   N+   N-   CNAME   VALUE   IC=VO
```

N+ is the positive node and N− is the negative node. The voltage of node N+ is assumed positive with respect to node N− and the current flows from node N+ through the capacitor to node N−. CNAME is the model name and VALUE is the nominal value of the capacitor. IC defines the initial (time-zero) voltage of the capacitor V_O.

TABLE 5.3   MODEL PARAMETERS FOR CAPACITORS

Name	Meaning	Units	Default
C	Capacitance multiplier		1
VC1	Linear voltage coefficient	Volt^{-1}	0
VC2	Quadratic voltage coefficient	Volt^{-2}	0
TC1	Linear temperature coefficient	°C^{-1}	0
TC2	Quadratic temperature coefficient	°C^{-2}	0

The model parameters are shown in Table 5.3. If CNAME is omitted, VALUE is the capacitance in farads and the VALUE can be positive or negative but *must* not be zero. If CNAME is included, the capacitance that depends on voltage and temperature is calculated from

```
CAP = VALUE * C * (1 + VC1 * V + VC2 * V²)
            * [1 + TC1 * (T - T0) + TC2 * (T - T0)²]
```

where T is the operating temperature in degrees Celsius and T0 is the room temperature in degrees Celsius.

Some Capacitor Statements

```
C1        6    5     10UF
CLOAD     12   11    5PF     IC=2.5V
CINPUT    15   14    ACAP    10PF
C2        20   19    ACAP    20NF    IC=1.5V
.MODEL    ACAP CAP (C=1  VC1=0.01  VC2=0.002  TC1=0.02  TC2=0.005)
```

Note. The initial condition (if any) applies only if the UIC (use initial condition) option is specified on the .TRAN command statement that is described in Sec. 6.9.

5.4.3 Inductor

The symbol for an inductor is L. The name of an inductor must start with L and it takes the general form of

```
L⟨name⟩ N+  N-  LNAME   VALUE   IC=I0
```

N+ is the positive node and N− is the negative node. The voltage of N+ is assumed positive with respect to node N− and the current flows from node N+ through the inductor to node N−. LNAME is the model name and VALUE is the nominal value of the inductor. IC defines the initial (time-zero) current of the inductor, I_0.

TABLE 5.4 MODEL PARAMETERS FOR INDUCTORS

Name	Meaning	Units	Default
L	Inductance multiplier		1
IL1	Linear current coefficient	Amps^{-1}	0
IL2	Quadratic current coefficient	Amps^{-2}	0
TC1	Linear temperature coefficient	°C^{-1}	0
TC2	Quadratic temperature coefficient	°C^{-2}	0

The model parameters of an inductor are shown in Table 5.4. If LNAME is omitted, VALUE is the inductance in henrys, and VALUE can be positive or negative but *must* not be zero. If LNAME is included, the inductance that depends on the current and temperature is calculated from

$$\text{IND} = \text{VALUE} * \text{L} * (1 + \text{IL1} * \text{I} + \text{IL2} * \text{I}^2)$$
$$* [1 + \text{TC1} * (\text{T} - \text{T0}) + \text{TC2} * (\text{T} - \text{T0})^2]$$

where T is the operating temperature in degrees Celsius and T0 is the room temperature in degrees Celsius.

Note. The initial condition (if any) applies only in the UIC (use initial condition) option is specified on the .TRAN command statement that is described in Section 6.9.

Some Inductor Statements

```
L1         6     5      10MH
LLOAD      12    11     5UH      IC=0.2MA
LLINE      15    14     LMOD     5MH
LCHOKE     20    19     LMOD     2UH      IC=0.5A
.MODEL  LMOD   IND (L=1   IL1=0.1   IL2=0.002   TC1=0.02   TC2=0.005)
```

Example 5.1

For the circuit in Figure 5.2(a), calculate and plot the transient response from 0 to 1 ms with a time increment of 5 μs. The output voltage is taken across resistor R_2 and the input voltage is shown in Figure 5.2(b). The results should be available for display and hard copy using *Probe*.

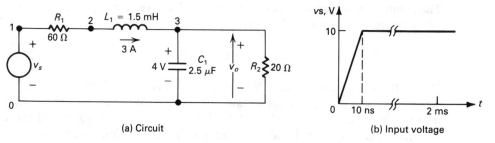

(a) Circuit (b) Input voltage

Figure 5.2 Circuit for Example 5.1

Solution   The circuit file contains the following statements.

Example 5.1   RLC Circuit
* The operating temperature is 50°
    .TEMP   50
* Input step voltage represented as a PWL waveform
    VS   1   0   PWL (0   0   10NS   10V   2MS   10V)
* R1 has a value of 60 Ohm with model RMOD
    R1   1   2   RMOD   60HM
* Inductor of 1.5mH with an initial current of 3 A and model name LMOD
    L1   2   3   LMOD   1.5MH   IC=3A
* Capacitor of 2.5UF with an initial voltage of 4 V and model name CMOD
    C1   3   0   CMOD   2.5UF   IC=4V
    R2   3   0   RMOD   20HM
* Model statements for resistor, inductor, and capacitor

```
.MODEL   RMOD   RES (R=1   TC1=0.02   TC2=0.005)
.MODEL   CMOD   CAP (C=1   VC1=0.01   VC2=0.002   TC1=0.02   TC2=0.005)
.MODEL   LMOD   IND (L=1   IL1=0.1   IL2=0.002   TC1=0.02   TC2=0.005)
```

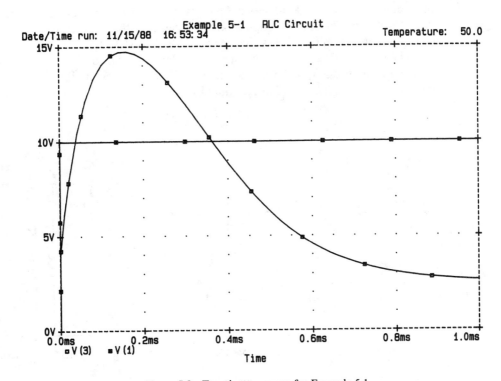

Figure 5.3 Transient response for Example 5.1

```
* Transient analysis from 0 to 1 ms with 5 μs time increment and using initial
* conditions (UIC).
  .TRAN  5US  1MS  UIC
* Plot the results of transient analysis − voltage at nodes 3
  .PLOT  TRAN  V(3)  V(1)
  .PROBE
  .END
```

The results of the simulation that are stored in output file EX5-1.OUT and
are obtained by *Probe* are shown in Figure 5.3.

5.5 MAGNETIC ELEMENTS

The magnetic elements are mutual inductors (transformers). The symbol for mutual coupling is *K*. The general form of coupled inductors is

```
K⟨name⟩  L⟨(1st inductor) name⟩  L⟨(2nd inductor) name⟩
+        ⟨(coupling) value⟩
```

For a couple inductor, K⟨name⟩ couples two or more inductors and coupling
⟨value⟩ is the coefficient of coupling *k*. The value of the coefficient of coupling
must be greater than 0 and less than or equal to 1, $0 < k \leq 1$.

The inductors can be coupled in either order. In terms of the *dot* convention
as shown in Figure 5.4(a), *PSpice* assumes a dot on the first node of each inductor. The mutual inductance is determined from

$$M = k\sqrt{L_1 L_2}$$

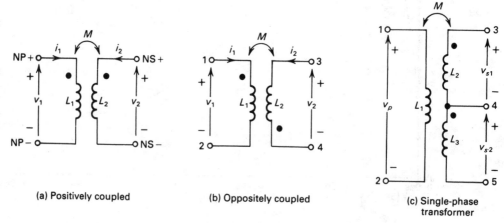

(a) Positively coupled   (b) Oppositely coupled   (c) Single-phase
                                                                    transformer

Figure 5.4 Coupled inductors

In time domain, the voltages of coupled inductors are expressed as

$$v_1 = L_1 \frac{di_1}{dt} + M \frac{di_2}{dt}$$

$$v_2 = M \frac{di_1}{dt} + L_2 \frac{di_2}{dt}$$

In frequency domain, the voltages are expressed as

$$V_1 = j\omega L_1 I_1 + j\omega M I_2$$

$$V_2 = j\omega M I_1 + j\omega L_2 I_2$$

where ω is the frequency in radians per second.

Some Coupled Inductor Statements

```
KTR       LA    LB    0.9
KIND      L1    L2    0.98
```

The coupled inductors in Figure 5.4(c) can be written as a single-phase transformer (with $k = 0.9999$)

```
*   PRIMARY
L1          1    2    0.5MH
*   SECONDARY
L2          3    4    0.5MH
*   MAGNETIC COUPLING
KXFRMER   L1   L2   0.9999
```

If the dot in the second coil is changed as shown in Figure 5.4(b), the coupled inductors are written as

```
L1          1    2    0.5MH
L2          4    3    0.5MH
KXFRMER   L1   L2   0.9999
```

A transformer with a single primary coil and center-tapped secondary, as shown in Figure 5.4(c), can be written as

```
*   PRIMARY
L1     1    2    0.5MH
*   SECONDARY
L2     3    4    0.5MH
L3     4    5    0.5MH
*   MAGNETIC COUPLING
K12   L1    L2   0.9999
K13   L1    L3   0.9999
K23   L2    L3   0.9999
```

These three statements can be written in *PSpice* as KALL L1 L2 L3 0.9999.

Notes

1. The name Kxx needs, not be related to the names of the inductors it is coupling. However, it is a good practice because it is convenient to identify the inductors involved in the coupling.
2. The polarity (or dot) is determined by the order of the nodes in the L... statements and not by the order of the inductors in the K... statement—e.g., (K12 L1 L2 0.9999) has the same result as (K12 L2 L1 0.9999).

For a nonlinear inductor, the general form is

```
K⟨name⟩  L⟨(inductor) name⟩  ⟨(coupling) value⟩
+        ⟨(model) name⟩ [(size) value]
```

For an iron-core transformer, k is very high and is greater than 0.999. The model type name for a nonlinear magnetic inductor is CORE; the model parameters are shown in Table 5.5. The [(size) value] scales the magnetic cross section and defaults to 1. It represents the number of lamination layers so that only one model statement can be used for a particular lamination type of core.

If the ⟨(model) name⟩ is specified, then the mutual coupling inductor becomes a nonlinear magnetic core and the inductors specify the number of turns instead of inductance. The list of the coupled inductors may be just one inductor. The magnetic core's B-H characteristics are analyzed using the Jiles-Atherton model [2]. The procedures to adjust the model parameters to specified B-H characteristics are described in Appendix C.

The statements for the coupled inductors in Figure 5.4(a) are similar to

```
*   Inductor L1 of 100 turns
L1          1    2    100
*   Inductor L2 of 10 turns
L2          3    4    10
*   Nonlinear coupled inductors with model CMOD
K12  L1  L2  0.9999  CMOD
*   Model for the nonlinear inductors
.MODEL CMOD CORE (AREA=2.0 PATH=62.8 GAP=0.1 PACK=0.98)
```

Note. A nonlinear magnetic model is not available in SPICE2.

Example 5.2

A circuit with two coupled inductors is shown in Figure 5.5. If the input voltage is 120 V peak, calculate the magnitude and phase of the output current for frequencies from 60 to 120 Hz with a linear increment. The total number of points in the sweep is 2. The coefficient of coupling for the transformer is 0.999.

Solution It is important to note that the primary and the secondary windings have a common node. Without this, *PSpice* will give an error message because there is no

TABLE 5.5 MODEL PARAMETERS FOR NONLINEAR MAGNETIC

Name	Meaning	Units	Default
AREA	Mean magnetic cross section	centimeters^2	0.1
PATH	Mean magnetic path length	centimeters	1.0
GAP	Effective air-gap length	centimeters	0
PACK	Pack (stacking)		1.0
MS	Magnetic saturation	A/meters	1E+6
ALPHA	Mean field parameter		1E−3
A	Shape parameter		1E+3
C	Domain wall-flexing constant		0.2
K	Domain wall-pinning constant		500

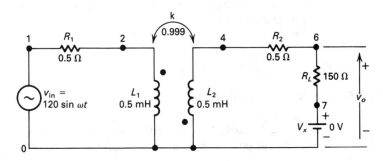

Figure 5.5 Circuit with two coupled inductors

dc path from the nodes of the secondary to the ground. The voltage source, $V_X = 0$ V, is connected to measure the output current I_L. The circuit file contains the following statements.

Example 5.2   Coupled Inductors

```
*    AC analysis where the frequency is varied linearly from
*    60 Hz to 120 Hz with 2 points.
.AC  LIN  2    60HZ  120HZ
*    Input voltage is 120 V peak and 0 degree phase for AC analysis.
VIN 1   0   AC   120
R1   1   2   0.5
*    The dot convention is followed in inductors L1 and L2.
L1   2   0   0.5MH
L2   0   4   0.5MH
*    Magnetic coupling coefficient is 0.999. The order of L1 and L2 is
*    not significant.
K12  L1   L2   0.999
R2   4   6   0.5
RL   6   7   150
```

```
*    A dummy voltage source of VX = 0 is added to measure the load current
VX   7   0   DC   0V
*    Print the magnitude and phase of output current. Some versions of
*    Pspice and SPICE do not permit reference to the currents through resistors,
*    e.g., IM(RL) IP(RL).
.PRINT  AC  IM(VX)  IP(VX)  IM(RL)  IP(RL)
.END
```

The results of the simulation that are stored in output file EX5-2.OUT are:

```
****       AC ANALYSIS                      TEMPERATURE =   27.000 DEG C
  FREQ         IM(VX)       IP(VX)      IM(RL)       IP(RL)
 6.000E+01    2.809E-01   -1.107E+02   2.809E-01   -1.107E+02
 1.200E+02    4.790E-01   -1.271E+02   4.790E-01   -1.271E+02
             JOB CONCLUDED
             TOTAL JOB TIME            8.18
```

5.6 LOSSLESS TRANSMISSION LINES

The symbol for a lossless transmission line is T. A transmission line has two ports—input and output. The general form of a transmission line is

```
T〈name〉  NA+  NA-  NB+  NB-  Z0=〈value〉  [TD=〈value〉]
+         [F=〈value〉  NL=〈value〉]
```

T〈name〉 is the name of the transmission line. NA+ and NA− are the nodes at the input port. NB+ and NB− are the nodes at the output port. NA+ and NB+ are defined as the positive nodes. NA− and NB− are defined as the negative nodes. The positive current flows from NA+ to NA− and from NB+ to NB−. Z_0 is the characteristic impedance.

The length of the line can be expressed in either of two forms: (1) the transmission delay TD may be specified, or (2) the frequency F may be specified together with NL, which is the normalized electrical length of the transmission line with respect to wavelength in the line at frequency F. If the frequency F is specified but NL is not, then the default value of NL is 0.25—that is, F has quarter-wave frequency. It should be noted that one of the options for expressing the length of the line must be specified—that is, TD or at least F must be specified. The block diagram of a transmission line is shown in Figure 5.6(a).

Some Transmission Statements

```
T1    1    2    3    4    Z0=50   TD=10NS
T2    4    5    6    7    Z0=50   F=2MHZ
TTRM  9   10   11   12    Z0=50   F=2MHZ   NL=0.4
```

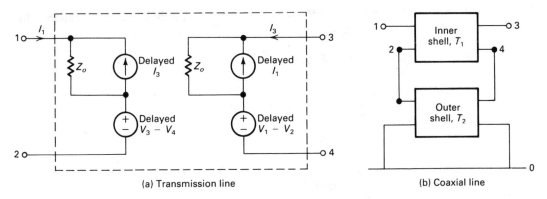

(a) Transmission line   (b) Coaxial line

Figure 5.6 Transmission line

A coaxial line, as shown in Figure 5.6(b), can be represented by two propagating lines, where the first line (T_1) models the inner conductor with respect to the shield and the second line (T_2) models the shield with respect to the outside. This is shown as follows:

```
T1    1    2    3    4    Z0=50    TD=1.5NS
T2    2    0    4    0    Z0=150   TD=1NS
```

Note. During the transient (.TRAN) analysis, the internal time step of *PSpice* is limited to be no more than one-half of the smallest transmission delay. Thus short transmission lines will cause long run times.

5.7 SWITCHES

PSpice allows simulation of a special kind of switch, as shown in Figure 5.7, whose resistance varies continuously depending on the voltage or current. When the switch is on, the resistance is R_{ON}, and when it is off, the resistance becomes R_{OFF}. Two types of switches are permitted in *PSpice:*

Voltage-controlled switch
Current-controlled switch

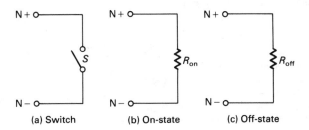

(a) Switch   (b) On-state   (c) Off-state

Figure 5.7 A switch with variable resistance.

Note. The voltage- and current-controlled switches are not available in SPICE2. However, they are available in SPICE3.

5.7.1 Voltage-Controlled Switch

The symbol for a voltage-controlled switch is S. The name of this switch must start with S and it takes the general form of

```
S⟨name⟩ N+  N-  NC+  NC-  SNAME
```

N+ and N− are the two nodes of the switch. The current is assumed to flow from N+ through the switch to node N−. NC+ and NC− are the positive and negative nodes of the controlling voltage source, as shown in Figure 5.8. SNAME is the model name. The resistance of the switch varies depending on the voltage across the switch. The type name for a voltage-controlled switch is VSWITCH, and the model parameters are shown in Table 5.6.

Voltage-Controlled Switch Statement

```
S1      6  5  4   0  SMOD
.MODEL  SMOD  VSWITCH (RON=0.5 ROFF=10E+6 VON=0.7 VOFF=0.0)
```

Notes

1. R_{ON} and R_{OFF} must be greater than zero and less than 1/GMIN. The value of GMIN can be defined as an option as described in .OPTIONS command in Section 6.5. The default value of conductance, GMIN is 1E-12 mhos.

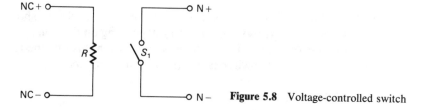

Figure 5.8   Voltage-controlled switch

TABLE 5.6   MODEL PARAMETERS FOR VOLTAGE-CONTROLLED SWITCH

Name	Meaning	Units	Default
VON	Control voltage for on-state	Volts	1.0
VOFF	Control voltage for off-state	Volts	0
RON	On resistance	Ohms	1.0
ROFF	Off resistance	Ohms	10^6

2. The ratio of R_{OFF} to R_{ON} should be less than 1E+12.

3. The difficulty due to high gain of an ideal switch can be minimized by choosing the value of R_{OFF} as high as permissible and that of R_{ON} as low as possible as compared to other circuit elements, within the limits of allowable accuracy.

Example 5.3

A circuit with a voltage-controlled switch is shown in Figure 5.9. If the input voltage is $v_s = 200 \sin(2000 \pi t)$, plot the voltage at node 3 and the current through the load resistor R_L for a time duration of 0 to 1 ms with an increment of 5 μs. The model parameters of the switch are RON=5M ROFF=10E+9 VON=25M VOFF=0.0. The results should be available for display by *Probe*.

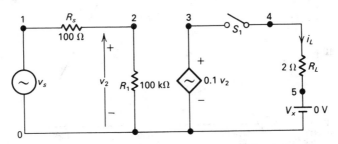

Figure 5.9   A circuit with voltage-controlled switch

Solution   The voltage source $V_X = 0$ V is inserted to monitor the output current. The listing of the circuit file is as follows.

```
    Example 5.3     Voltage-Controlled Switch
*    Sinusoidal input voltage of 200 V peak with 0 phase delay
VS   1   0   SIN (0   200V   1KHZ)
RS   1   2   100OHM
R1   2   0   100KOHM
*    Voltage-controlled voltage source with a voltage gain of 0.1
E1   3   0   2   0   0.1
RL   4   5   2OHM
*    A dummy voltage source of VX = 0 to measure the load current
VX   5   0   DC   0V
*    Voltage-controlled switch controlled by voltage across nodes 3 and 0
S1   3   4   3   0   SMOD
*    Switch model descriptions
.MODEL   SMOD   VSWITCH (RON=5M ROFF=10E+9 VON=25M   VOFF=0.0)
*    Transient analysis from 0 to 1 ms with 5-µs increment
.TRAN   5US   1MS
*    Plot the current through VX and the input voltage.
.PLOT   TRAN   I(VX) V(3)
.PROBE
.END
```

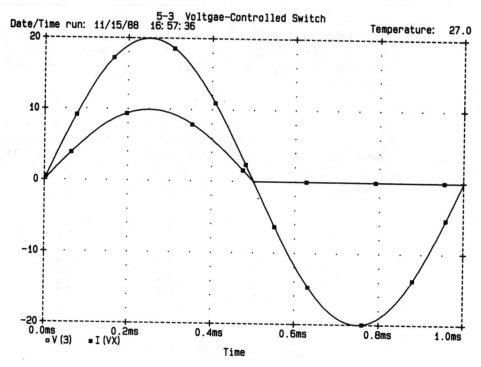

Figure 5.10 Transient response for Example 5.3

The results of the simulation that are stored in output file EX5-3.OUT and obtained by *Probe* are shown in Figure 5.10.

5.7.2 Current-Controlled Switch

The symbol for a current-controlled switch is *W*. The name of the switch must
start with *W* and it takes the general form of

```
W⟨name⟩ N+    N−   VN    WNAME
```

N+ and N− are the two nodes of the switch. V_N is a voltage source through
which the controlling current flows, as shown in Figure 5.11. WNAME is the
model name. The resistance of the switch depends on the current through the
switch. The type name for a current-controlled switch is ISWITCH, and the
model parameters are shown in Table 5.7.

Current-Controlled Switch Statement

```
W1        6 5  VN  RELAY
.MODEL   RELAY ISWITCH (RON=0.5 ROFF=10E+6 ION=0.07 IOFF=0.0)
```

TABLE 5.7 MODEL PARAMETERS FOR
CURRENT-CONTROLLED SWITCH

Name	Meaning	Units	Default
ION	Control current for on-state	Amps	1E-3
IOFF	Control current for off-state	Amps	0
RON	On resistance	Ohms	1.0
ROFF	Off resistance	Ohms	10^6

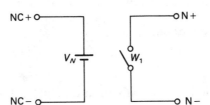

Figure 5.11 Current-Controlled Switch

Note. The current through voltage source V_N controls the switch. The voltage source V_N must be an independent source, and it can have a zero or finite value. The limitations of the parameters are similar to that for the voltage-controlled switch.

Example 5.4

A circuit with a current-controlled switch is shown in Figure 5.12. Plot the capacitor voltage and the current through the inductor for a time duration of 0 to 160 μs with an increment of 1 μs. The model parameters of the switch are: RON=1E+6 ROFF=0.001 ION=1MA IOFF=0. The results should be available for display by *Probe*.

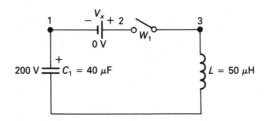

Figure 5.12 Circuit with current-controlled switch

Solution The voltage source $V_X = 0$ V is inserted to monitor the controlling current. The listing of the circuit file is as follows.

```
Example 5.4    Current-Controlled Switch
*   C1 of 40 UF with an initial voltage of 200 V
C1   1   0   40UF   IC=200
```

```
*    Dummy voltage source of VX=0
VX   2   1   DC   0V
*    Current-controlled switch with model name SMOD
W1   2   3   VX   SMOD
*    Model parameters
.MODEL   SMOD   ISWITCH (RON=1E+6 ROFF=0.001 ION=1MA IOFF=0)
L1   3   0   50UF
*    Transient analysis with UIC (use initial condition)
option
.TRAN   1US   160US   UIC
*    Plot the voltage at node 1 and the current through VX.
.PLOT TRAN   V(1)   I(VX)
*    Graphic post processor
.PROBE
.END
```

The results of the simulation that are stored in output file EX5-4.OUT and obtained by *Probe* are shown in Figure 5.13.

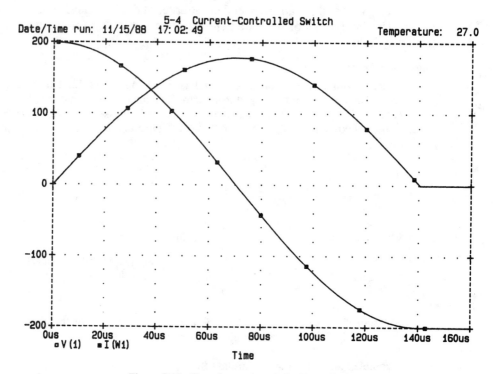

Figure 5.13   Transient response for Example 5.4

SUMMARY

The symbols for the passive elements are:

C   Capacitor
         C⟨name⟩ N+ N− CNAME VALUE IC=VO

L   Inductor
         L⟨name⟩ N+ N− LNAME IC=IO

K   Mutual inductors (Transformers)
         K⟨name⟩ L⟨(1st inductor) name⟩ L⟨(2nd inductor)name⟩ ⟨value)]

K   Nonlinear inductor
         K⟨name⟩ L⟨(inductor) name⟩ ⟨(coupling) value⟩
         + ⟨(model) name⟩ [(size) value]

R   Resistor
         R⟨name⟩ N+ N− RNAME VALUE

S   Voltage-controlled switch
         S⟨name⟩ N+ N− NC+ NC− SNAME

T   Lossless transmission lines
         T⟨name⟩ NA+ NA− NB+ NB− ZO=⟨value⟩ [TD=⟨value⟩]
         + [F=⟨value⟩ NL=⟨value⟩]

W   Current-controlled switch
         W⟨name⟩ N+ N− VN WNAME

REFERENCES

1. *PSpice Manual.* Irvine, Calif.: MicroSim Corporation, 1988.
2. D. C. Jiles and D. L. Atherton, "Theory of ferromagnetic hysteresis," *Journal of Magnetism and Magnetic Material,* Vol. 61, No. 48, 1986, pp. 48–60.

PROBLEMS

Write the *PSpice* statements for the following problems:

5.1. A resistor R_1, is connected between nodes 3 and 4, and has a nominal value of $R = 10$ kΩ. The operating temperature is 55°C and it has the form

$$R_1 = R * [1 + 0.2 * (T - T0) + 0.002 * (T - T0)^2]$$

5.2. A resistor R_1, is connected between nodes 3 and 4, and has a nominal value of $R = 10$ kΩ. The operating temperature is 55°C and it has the form

```
    4.5*(T - TO)
R₁ = R * 1.01
```

5.3. A capacitor C_1, is connected between nodes 5 and 6, and has a value of 10 pF and an initial voltage of -20 V.

5.4. A capacitor C_1, is connected between nodes 5 and 6, and has a nominal value of $C = 10$ pF. The operating temperature is $T = 55$°C. The capacitance, which is a function of its voltage and the operating temperature, is given by

```
C₁ = C * (1 + 0.01 * V + 0.002 * V²)
       * [1 + 0.03 * (T - TO) + 0.05 * (T - TO)²]
```

5.5. An inductor L_1, is connected between nodes 5 and 6, and has a value of 0.5 mH and carries an initial current of 0.04 mA.

5.6. An inductor L_1, is connected between nodes 3 and 4, and has a nominal value of $L = 1.5$ mH. The operating temperature is $T = 55$°C. The inductance, which is a function of its current and the operating temperature, is given by

```
L₁ = L * (1 + 0.01 * I + 0.002 * I²)
       * [1 + 0.03 * (T - TO) + 0.05 * (T - TO)²]
```

5.7. The two inductors that are oppositely coupled as shown in Figure 6.4(b) are $L_1 = 1.2$ mH and $L_2 = 0.5$ mH. The coefficients of coupling are $K_{12} = K_{21} = 0.999$.

5.8. Plot the transient response of the circuit in Figure P5.8 from 0 to 5 ms with a time increment of 25 μs. The output voltage is taken across the across the capacitor. Use *Probe* for graphical output.

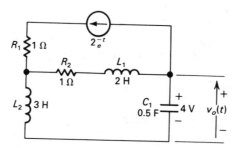

Figure P5.8

5.9. Repeat Problem 5.8 for the circuit in Figure P5.9.

5.10. Plot the frequency response of the circuit in Figure P5.10 from 10 Hz to 100 kHz with a decade increment and 10 points per decade. The output voltage is taken across the capacitor. Print and plot the magnitude and phase angle of the output voltage. Assume a source voltage of 1 V peak.

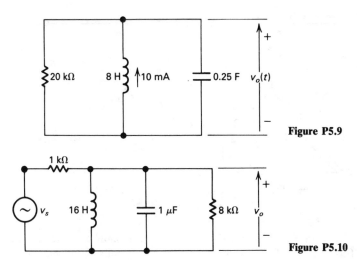

Figure P5.9

Figure P5.10

5.11. A single-phase transformer, as shown in Figure P5.11, has a center-tapped primary, where $L_p = 1.5$ mH, $L_s = 1.3$ mH, and $K_{ps} = K_{sp} = 0.999$.

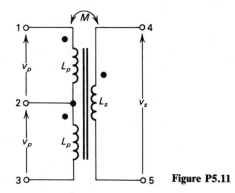

Figure P5.11

5.12. A three-phase transformer, which is shown in Figure P5.12, has $L_1 = L_2 = L_3 = 1.2$ mH and $L_4 = L_5 = L_6 = 0.5$ mH. The coupling coefficients between the primary and secondary windings of each phase are $K_{14} = K_{41} = K_{25} = K_{52} = K_{36} = K_{63} = 0.9999$. There is no cross coupling with other phases.

5.13. A switch, which is connected between nodes 5 and 4, is controlled by a voltage source between nodes 3 and 0. The switch will conduct if the controlling voltage is 0.5 V. The on-state resistance is 0.5 Ω and the off-state resistance is 2E+6 Ω.

5.14. A switch, which is connected between nodes 5 and 4, is controlled by a current. The voltage source V_1 through which the controlling current flows is connected between nodes 2 and 0. The switch will conduct if the controlling current is 0.55 mA. The on-state resistance is 0.5 Ω and the off-state resistance is 2E+6 Ω.

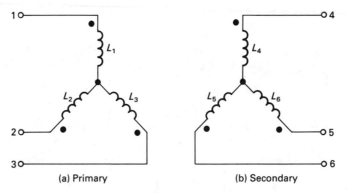

(a) Primary   (b) Secondary

Figure P5.12

5.15. For the circuit in Figure P5.15, plot the transient response of the load and source current for 5 cycles of the switching period with a time increment of 25 μs. The model parameters of the voltage-controlled switches are: RON=0.025, ROFF=1E+8, VON=0.05 and VOFF=0. The output should also be available for display and hard copy by *Probe*.

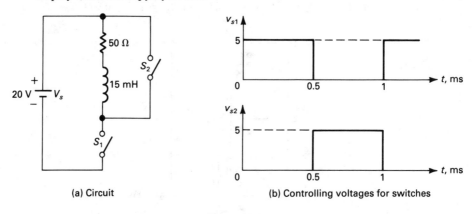

(a) Circuit   (b) Controlling voltages for switches

Figure P5.15

6

Dot Commands

6.1 INTRODUCTION

PSpice has various commands for performing various analyses, getting different types of output and modeling elements. These commands begin with a dot and are known as *dot commands*. These commands can be used to specify

Models
Types of output
Operating temperature and end of circuit
Options
Dc analysis
Ac analysis
Noise analysis
Transient analysis
Fourier analysis

Note. If you are not sure about a command and its effect, run a circuit file with the command and then look at the results.

6.2 MODEL

PSpice allows one (1) to model an element based on its parameters, (2) to model a small circuit that is repeated a number of times in the main circuit, and (3) to use a model that is defined in another file. The commands are

1. .MODEL   Model
2. .SUBCKT   Subcircuit
3. .ENDS   End of subcircuit
4. .LIB   Library file
5. .INC   Include file

6.2.1 .MODEL Model

The .MODEL command has already been discussed in Section 5.2.

6.2.2 .SUBCKT Subcircuit

A **subcircuit** permits one to define a block of circuitry and then to use that block in several places. The general form for subcircuit definition (or description) is

```
.SUBCKT  SUBNAME  [(two or more) nodes]
```

The symbol for a subcircuit call is *X*. The general form of a call statement is

```
X(name)  [(two or more) nodes]   SUBNAME
```

SUBNAME is the name of the subcircuit definition and ⟨(two or more) nodes⟩ are the nodes of the subcircuit. X⟨name⟩ causes the referenced subcircuit to be inserted into the circuit with given nodes replacing the argument nodes in the definition. The subcircuit name, SUBNAME, may be considered as equivalent to a subroutine name in FORTRAN programing, where X⟨name⟩ is the call statement and [nodes] are the variables or arguments of the subroutine.

Subcircuits may be nested. That is, subcircuit *A* may call other subcircuits. But the nesting cannot be circular, which means that if subcircuit *A* contains a call to subcircuit *B*, then subcircuit *B* must not contain a call to subcircuit *A*.

There must be the same number of nodes in the subcircuit calling statement as in its definition. The subcircuit definition should contain only element statements (statements without a dot) and may contain .MODEL statements.

6.2.3 .ENDS End of Subcircuit

A subcircuit must end with an .ENDS statement. The end of a subcircuit definition has the general form

```
.ENDS  SUBNAME
```

SUBNAME is the name of the subcircuit and it indicates which subcircuit description is to be terminated. If the .ENDS statement is missing, all subcircuit descriptions are terminated.

End of Subcircuit Statements

```
.ENDS OPAMP
.ENDS
```

Note. The name of the subcircuit can be omitted. However, it is advisable to identify the name of the subcircuit to be terminated, especially if there is more than one subcircuit.

Example 6.1

Write the subcircuit call and subcircuit description for the circuit in Figure 6.1.

Solution The list of statements for subcircuit call and description is as follows.

```
*    The call statement X1 is connected to nodes 1 and 2.  The subcircuit
name is EQVT.  Nodes 1 and 2 are referred to the main circuit file.
X1    1    0    EQVT
*    The subcircuit definition.  Nodes 1 and 3 are referred to the subcircuit
.SUBCKT EQVT  1    3
*    Sinusoidal voltage of 0.5 V at 1 kHz
V1    1    3    SIN (0   0.5    1KHZ)
R1    1    2    5K
R2    2    3    2k
C1    2    3    0.1UF
*    End of subcircuit definition
.ENDS  EQVT
```

Note. There is no interaction between the nodes in the main circuit and the subcircuit. Node numbers in the subcircuit are independent of those in the main circuit. However, the subcircuit should not have node 0 because node 0, which is considered global by *PSpice,* is the ground.

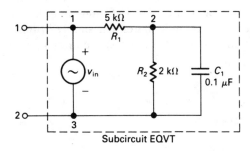

Subcircuit EQVT

Figure 6.1 Subcircuit

6.2.4 .LIB Library File

A library file may be referenced into the circuit file by using the statement

```
.LIB   FNAME
```

FNAME is the name of the library file to be called. A library file may contain: comments, .MODEL statements, subcircuit definition, .LIB statements and .END statements. No other statements are permitted. If FNAME is omitted, *PSpice* looks for the default file, EVAL.LIB, that comes with *PSpice* programs. The library file FNAME may call for another library file.

When a .LIB command calls for a file, it does not bring the whole text of the library file into the circuit file. It simply reads those models or subcircuits that are called by the main circuit file. As a result, only those models or subcircuit descriptions that are needed by the main circuit file take up the main memory (RAM) space.

Note. Check the PSpice files for default library file name.

Some Library File Statements

```
.LIB
.LIB   DNOM.LIB   (Library file DNOM.LIB is on the default drive and
                    describes the parameters of diode model).
.LIB   B:\LIB\MNOM.LIB (Library file MNOM.LIB is on directory file LIB
                 in drive B: and describes the parameters of MOSFET model).
.LIB   C:\LIB\MNOM.LIB (Library file MNOM.LIB is in directory file LIB
                 on drive C: and describes the parameters of MOSFET model).
```

6.2.5 .INC Include File

The contents of another file may be included into the circuit file using the statement

```
.INC   NFILE
```

NFILE is the name of the file to be included and can be any character string that is a legal file name for computer systems. It may include a volume, directory, and version number.

Included files may contain any statements except a title line. However, a comment line may be used instead of a title line. If an .END statement is present, it marks only the end of the included file. An .INC statement may be used up to only 4 levels of "including." The include statement simply brings everything in the included file into the circuit file and takes up space in main memory (RAM).

Some Include File Statements

```
.INC   OPAMP.CIR
.INC   a:INVETER.CIR
.INC   c:\LIB\NOR.CIR
```

6.3 TYPES OF OUTPUT

The commands that are available to get output from the results of simulations are

```
.PRINT    Print
.PLOT     Plot
.PROBE    Probe
 Probe Output
.WIDTH    Width
```

6.3.1 .PRINT Print

The results from DC, AC, transient (TRAN), and noise (NOISE) analyses can be obtained in the form of tables. The print statement takes one of the forms

```
.PRINT   DC     [output variables]
.PRINT   AC     [output variables]
.PRINT   TRAN   [output variables]
.PRINT   NOISE  [output variables]
```

The maximum number of output variables is 8 in any .PRINT statement. However, more than one .PRINT statement can be used to print all the desired output variables.

The values of the output variables are printed as a table with each column corresponding to one output variable. The number of digits for output values can be changed by the NUMDGT option on the .OPTIONS statement in Section 6.5. The results of the .PRINT statement are stored in the output file.

Some Print Statements

```
.PRINT   DC V(2), V(3,5), V(R1), VCE(Q2), I(VIN), I(R1), IC(Q2)
.PRINT   AC VM(2), VP(2), VM(3,5), V(R1), VG(5), VDB(5), IR(5), II(5)
.PRINT   NOISE INOISE ONOISE DB(INOISE) DB(ONOISE)
.PRINT   TRAN V(5) V(4,7) (0,10V) IB(Q1) (0,50MA) IC(Q1) (-50MA, 50MA)
```

Note. Having two .PRINT statements for the same variables will not produce two tables. *PSpice* will ignore the first statement and produce output for the second statement.

6.3.2 .PLOT Plot

The results from DC, AC, transient (TRAN), and noise (NOISE) analyses can be obtained in the form of line printer plots. The plots are drawn by using characters, and the results can be obtained from any kind of printer. The plot statement takes one of the following forms:

```
.PLOT  DC    [output variables]
          + (<(lower limit) value), <(upper limit) value))
.PLOT  AC    [output variables]
          + [<(lower limit) value), <(upper limit) value)]
.PLOT TRAN   [output variables]
          + [<(lower limit) value), <(upper limit) value)]
.PLOT NOISE  [output variables]
          + [<(lower limit) value), <(upper limit) value)]
```

The maximum number of output variables is 8 in any .PLOT statement. More than one .PLOT statement can be used to plot all the desired output variables.

The range and increment of the x-axis is fixed by the type of analysis command (e.g., .DC or .AC or .TRAN or .NOISE). The range of y-axis is set by adding (<(lower limit) value), <(upper limit) value)) at the end of .PLOT statement. The y-axis range, (<(lower limit) value), <(upper limit) value)) can be placed in the middle of a set of output variables. The output variables will follow the specified range that comes immediately to the right.

If the y-axis range is omitted, *PSpice* assigns a default range determined by the range of the output variable. If the ranges of output variables vary widely, *PSpice* assigns the ranges corresponding to the different output variables.

Some Plot Statements

```
.PLOT DC V(2), V(3,5), V(R1), VCE(Q2), I(VIN), I(R1), IC(Q2)
.PLOT AC VM(2), VP(2), VM(3,5), V(R1), VG(5), VDB(5), IR(5), II(5)
.PLOT NOISE INOISE ONOISE DB(INOISE) DB(ONOISE)
.PLOT TRAN V(5) V(4,7) (0,10V) IB(Q1) (0, 50MA) IC(Q1) (-50MA, 50MA)
```

Notes. In the first three statements, the y-axis is by default. In the last statement, the range for voltages V(5) and V(4,7) is 0 V to 10 V, that for current IB(Q1) is 0 MA to 50 MA, and that for the current IC(Q1) is −50 MA to 50 MA.

6.3.3 .PROBE *Probe*

Probe is a graphics post-processor and is available as an option for the professional version of *PSpice*. However, *Probe* comes with the student version of *PSpice*. The results from the DC, AC, and transient (TRAN) analysis cannot be used directly by *Probe*. First, the results have to be processed by the .PROBE command, which writes the processed data on a file, PROBE.DAT, for use by the *Probe*. The command takes one of these forms:

```
.PROBE
.PROBE  <one or more output variables>
```

In the first form, where no output variable is specified, the .PROBE command writes all the node voltages and all the element currents into the

PROBE.DAT file. The element currents are written in the forms that are permitted as output variables and are discussed in Section 3.3.2.

In the second form, where the output variables are specified, *PSpice* writes only the specified output variables to the PROBE.DAT file. This form is suitable for users without a fixed disk to limit the size of the PROBE.DAT file.

Probe Statements

```
.PROBE
.PROBE V(5), V(4,3), V(C1), VM(2), I(R2), IB(Q1), VBE(Q1)
```

6.3.4 *Probe* Output

It is very easy to use *Probe*. Once the results of the simulations are processed by the .PROBE command, the results are available for graphics output. *Probe* comes with a first menu, as shown in Figure 6.2, to choose the type of analysis. After the first choice, the second level is the choice for the plots and coordinates of output variables, as shown in Figure 6.3. After the choices, the output is displayed as shown in Figure 6.4.

```
                              ┌──────┐
                              │ Probe │
                              └──────┘

              Graphics Post-Processor for PSpice
                   Version 1.13 - October 1987
           © Copyright 1985, 1986, 1987 by MicroSim Corporation
                            --------
                       Classroom Version
           Copying of this program is welcomed and encouraged


          Circuit:  EXAMPLE1 - An Illustration of all Commands
        Date/Time run:  10/31/88 16:15:19          Temperature: 35.0




   0)Exit Program 1)DC Sweep 2)AC Sweep 3)Transient Analysis :   1
```

Figure 6.2   Select analysis display for *Probe*

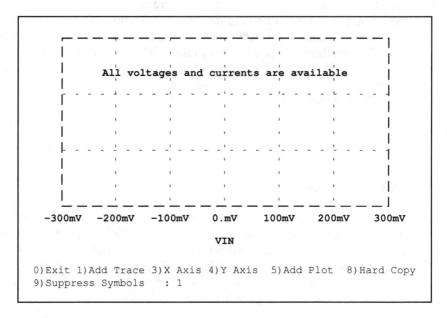

Figure 6.3 Select plot/graphics output

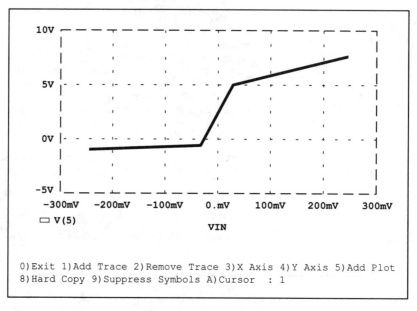

Figure 6.4 Output display

With one exception, *Probe* disregards upper- and lowercase: V(4) and v(4) are equivalent. The exception is the prefixes M and m: m means *milli* (1E−3), whereas M means *mega* (1E+6).

The suffixes MEG and MIL are not available. The units that are recognized by *Probe* are

V	Volts
A	amps
W	watts
d	degrees (of phase)
s	seconds
H	hertz

Probe also recognizes that W = V * A, V = W/A, and A = W/V. Therefore, the addition of a trace, that is,

```
VCE(Q1)*IC(Q1)
```

gives the power dissipation of transistor Q_1 and will be labeled with W. Arithmetic expressions of output variables are also allowed; the available operators are: +, −, *, /, and parentheses. The available functions are

FUNCTION	MEANING
ABS(x)	$|x|$ (absolute value)
B(Kxy)	Flux density of coupled inductor Kxy
H(Kxy)	Magnetization of coupled inductor Kxy
SGN(x)	+1 (if $x > 0$), 0 (if $x = 0$), −1 (if $x < 0$)
EXP(x)	e^x
DB(x)	$20*log$ ($|x|$) (base 10 log)
LOG(x)	$ln(x)$ (base *e* log)
LOG10(x)	$log(x)$ (base 10 log)
PWR(x,y)	$|x|^y$
SQRT(x)	$x^{1/2}$
SIN(x)	$sin(x)$ (x in radians)
COS(x)	$cos(x)$ (x in radians)
TAN(x)	$tan(x)$ (x in radians)
ARCTAN(x)	$tan^{-1}(x)$ (result in radians)
d(y)	Derivative of y with respect to the x-axis variable
s(y)	Integral of y over the x-axis variable
AVG(x)	Running average of x
RMS(x)	Running RMS average of x

For derivatives and integrals of simple variables (not expressions), the shorthand notations that are available are

dV(4) is equivalent to d(V(4)).
sIC(Q3) is equivalent to s(IC(Q3)).

The plot of

```
dIC(Q2)/dIB(Q2)
```

will give the small-signal beta of Q2.

Two or more traces can be added with only one Add Trace command, where all the expressions are separated by a space or a comma. For instance,

```
V(2)  V(4), IC(M1), RMS(I(VIN))
```

adds four traces. This gives the same result as using Add Trace four times with only one trace at a time but is faster, since the plot is not redrawn between adding each trace.

The PROBE.DAT file can contain more than one of any kind (e.g., two transient analyses with three temperatures). If the *PSpice* run is for a transient analysis at three temperatures, the expression

```
V(1)
```

will result in *Probe* drawing three curves instead of the usual one curve.

Entering the expression

```
V(1)@n
```

will result in drawing the curve of V(1) for the *n*th transient analysis.

Entering the expression

```
V(1)@1-V(1)@2
```

will display the difference between the waveforms from the first and second temperature, whereas the expression

```
V(1)-V(2)@2
```

will display three curves, one for each V(1).

Notes

1. The .PROBE command requires a math coprocessor for the professional version of *PSpice*, but it is not required for the student version.
2. *Probe* is not available on SPICE. However, the newest version of SPICE (SPICE3) has a post-processor similar to *Probe*, called *Nutmeg*.
3. It is required that the type of display and the type of hard-copy devices on the PROBE.DEV file be specified as follows:

```
Display = ⟨display name⟩
Hard copy = ⟨port name⟩, ⟨device name⟩
```

The details of names for display, port, and device (printer) can be found in the README.DOC file that comes with the *PSpice* programs or in the *PSpice* manual.

4. The display and hard-copy devices can be set from the "Display/Printer Setup" menu.

6.3.5 .WIDTH Width

The width of the output in columns can be set by the .WIDTH statement, which has the general form of

```
.WIDTH  OUT=⟨value⟩
```

The ⟨value⟩ is in columns and must be either 80 or 132. The default value is 80.

6.4 TEMPERATURE AND END OF CIRCUIT

The temperature command is discussed in Section 4.2. The last statement for the end of a circuit is

```
.END
```

All data and commands must come before the .END command. The .END command instructs *PSpice* to perform all the specified analysis of the circuit; after processing the results of the specified analysis, *PSpice* resets itself for performing any further computations.

Note. An input file may have more than one circuit, where each circuit has its .END command. *PSpice* will perform all the specified analysis and processes the results of each circuit one by one. *PSpice* resets everything at the beginning of each circuit. Instead of running *PSpice* separately for each circuit, this is a convenient way for performing the analysis of many circuits with one run statement.

6.5 OPTIONS

PSpice allows various options to control and to limit parameters for the various analysis. The general form is

```
.OPTIONS [(options) name)] [((options) name)=(value)]
```

The options can be listed in any order. There are two types of options: (1) those without values and (2) those with values. The options without values are used as

TABLE 6.1   LIST OF OPTIONS WITHOUT VALUES

Option	Effects
NOPAGE	Suppresses paging and printing of a banner for each major section of output
NOECHO	Suppresses listing of the input file
NODE	Causes output of net list (node table)
NOMOD	Suppresses listing of model parameters
LIST	Causes summary of all circuit elements (or devices) to be output
OPTS	Causes values for all options to be output
ACCT	Summary and accounting information is output at the end of all the analysis.
WIDTH	Same as .WIDTH OUT= statement

TABLE 6.2   LIST OF OPTIONS WITH VALUES

Option	Effects	Units	Default
DEFL	MOSFET channel length (L)	Meter	100u
DEFW	MOSFET channel width (W)	Meter	100u
DEFAD	MOSFET drain diffusion area (AD)	Meter^{-2}	0
DEFAS	MOSFET source diffusion area (AS)	Meter^{-2}	0
TNOM	Default temperature (also the temperature at which model parameters are assumed to have been measured)	Degrees Celsius	27
NUMDGT	Number of digits output in print tables		4
CPTIME	CPU time allowed for a run	Second	1E6
LIMPTS	Maximum points allowed for any print table or plot		201
ITL1	DC and bias point "blind" iteration limit		40
ITL2	DC and bias point "educated guess" iteration limit		20
ITL4	Iteration limit at any point in transient analysis		10
ITL5	Total iteration limit for all points in transient analysis (ITL5=0 means ITL5=infinite)		5000
RELTOL	Relative accuracy of voltages and currents		0.001
TRTOL	Transient analysis accuracy adjustment		7.0
ABSTOL	Best accuracy of currents	Amp	1pA
CHGTOL	Best accuracy of charges	Coulomb	0.01pC
VNTOL	Best accuracy of voltages	Volt	1uV
PIVREL	Relative magnitude required for pivot in matrix solution		1E-13
GMIN	Minimum conductance used for any branch	Ohm^{-1}	1E-12

flags of various kinds and only the option name is mentioned. Table 6.1 shows the options without values.

The options with values are used to specify certain optional parameters. The option names and their values are specified. Table 6.2 shows the options with values. The commonly used options are: NOPAGE, NOECHO, NOMOD, TNOM, CPTIME, NUMDGT, GMIN, and LIMPTS.

Options Statements

```
.OPTIONS    NOPAGE  NOECHO  NOMOD  DEFL=20U  DEFW=15U  DEFAD=50P  DEFAS=50P
.OPTIONS    ACCT  LIST  RELTOL=.005
```

Job Statistics Summary

If the option ACCT is specified on the .OPTIONS statement, *PSpice* will print various statistics about the run at the end. This option is not normally required for most circuit simulations. This list follows the format of the output.

ITEM	MEANING
NUNODS	Number of distinct circuit nodes before subcircuit expansion.
NCNODS	Number of distinct circuit nodes after subcircuit expansion. If there are no subcircuits, then NCNODS = NUNODS.
NUMNOD	Total number of distinct nodes in circuit. This is NCNODS plus the internal nodes generated by parasitic resistances. If no device has parasitic resistances, then NUMNOD = NCNODS.
NUMEL	Total number of devices (or elements) in circuit after subcircuit expansion. This includes all statements that do not begin with . or X.
DIODES	Number of diodes after subcircuit expansion.
BJTS	Number of bipolar transistors after subcircuit expansion.
JFETS	Number of junction FETs after subcircuit expansion.
MFETS	Number of MOSFETs after subcircuit expansion.
GASFETS	Number of GaAs MESFETs after subcircuit expansion.
NUMTEM	Number of different temperatures.
ICVFLG	Number of steps of DC sweep.
JTRFLG	Number of print steps of transient analysis.

JACFLG	Number of steps of AC analysis.
INOISE	1 or 0: Noise analysis was/was not done.
NOGO	1 or 0: Run did/did not have an error.
NSTOP	The circuit matrix is conceptually (not physically) of dimension NSTOP × NSTOP.
NTTAR	Actual number of entries in circuit matrix at beginning of run.
NTTBR	Actual number of entries in circuit matrix at end of run.
NTTOV	Number of terms in circuit matrix that come from more than one device.
IFILL	Difference between NTTAR and NTTBR.
IOPS	Number of floating-point operations needed to do one solution of circuit matrix.
PERSPA	Percent sparsity of circuit matrix.
NUMTTP	Number of internal time steps in transient analysis.
NUMRTP	Number of times in transient analysis that a time step was too large and had to be cut back.
NUMNIT	Total number of iterations for transient analysis.
MEMUSE/MAXMEM	Amount of circuit memory used/available in bytes. There are two memory pools. Exceeding either one will abort the run.
COPYKNT	Number of bytes that were copied in the course of doing memory management for this run.
READIN	Time spent reading and error checking the input file.
SETUP	Time spent setting up the circuit matrix pointer structure.
DCSWEEP	Time spent and iteration count for calculating DC sweep.
BIASPNT	Time spent and iteration count for calculating bias point and bias point for transient analysis.
MATSOL	Time spent solving circuit matrix (this time is also included in each analysis' time). The iteration count is the number of times the rows or columns were swapped in the course of solving it.
ACAN	Time spent and iteration count for AC analysis.

TRANAN	Time spent and iteration count for transient analysis.
OUTPUT	Time spent preparing .PRINT tables and .PLOT plots.
LOAD	Time spent evaluating device equations (this time is also included in each analysis time).
OVERHEAD	Other time spent during run.
TOTAL JOB TIME	Total run time excluding the time to load the program files PSPICE1.EXE and PSPICE2.EXE into memory.

6.6 DC ANALYSIS

In DC analysis, all the independent and dependent sources are of DC types. The inductors and capacitors in a circuit are considered as short circuits and open circuits, respectively. This is due to the fact that at zero frequency, the impedance represented by an inductor is zero and that by a capacitor is infinite. The commands that are available for DC analyses are

.OP	DC operating point
.NODESET	Nodeset
.SENS	Small-signal sensitivity
.TF	Small-signal transfer function
.DC	DC sweep

6.6.1 .OP Operating Point

Electronic circuits contain nonlinear devices (e.g., diodes, transistors) whose parameters depend on the **operating point.** The operating point is also known as a **bias point** or **quiescent point.** The operating point is always calculated by *PSpice* for calculating the small-signal parameters of nonlinear devices during the DC sweep and transfer function analysis. The command takes the form

    .OP

The .OP command controls the output of the bias point but not the method of bias analysis and the results of bias point. If the .OP command is omitted, *PSpice* prints only a list of the node voltages. If the .OP command is present, *PSpice* prints the currents and power dissipations of all the voltages. The small-signal parameters of all nonlinear controlled sources and all the semiconductor devices are also printed.

6.6.2 .NODESET Nodeset

In calculating the operating bias point, some or all of the nodes of the circuit may be assigned initial guesses to help DC convergence by the statement, as in

```
.NODESET  V(1)=V1 V(2)=V2 ··· V(N)=VN
```

$V(1)$, $V(2)$, $\cdot\ \cdot\ \cdot$ at the node voltages and V_1, V_2, $\cdot\ \cdot\ \cdot$ are their respective values of the initial guesses. Once the operating point is found, the .NODESET command has no effect during the DC sweep or transient analysis. This command may be necessary for convergence on flip-flop circuits to "break the tie-in" condition. In general, this command should not be necessary. One should not confuse it with the .IC command, which sets the initial conditions of the circuits during the operating point calculations for transient analysis. The .IC command is discussed in Section 6.9.1.

Statement for Nodeset

```
.NODESET  V(4)=1.5V  V(6)=0  V(25)=1.5V
```

6.6.3 .SENS Sensitivity Analysis

The sensitivity of output voltages or currents with respect to each and every circuit and device parameter can be calculated by the .SENS statement, which has the general form of

```
.SENS  ((one or more output variables)
```

The .SENS statement calculates the bias point and the linearized parameters around the bias point. In this analysis, the inductors are assumed to be short circuits, and capacitors are assumed to be open circuits. If the output variable is a current, then that current must be through a voltage source. The sensitivity of each output variable with respect to all the device values and model parameters are calculated, and the .SENS statement prints the results automatically. Therefore, it should be noted that a .SENS statement may generate a huge amount of data if many output variables are specified.

Statement for Sensitivity Analysis

```
.SENS  V(5), V(2.3) I(V2) I(V5)
```

Example 6.2

For the circuit in Figure 6.5(a), calculate and print the sensitivity of output voltage V(3) with respect to each circuit element. The amplifier is represented by the subcircuit in Figure 6.5(b). The operating temperature is 40°C.

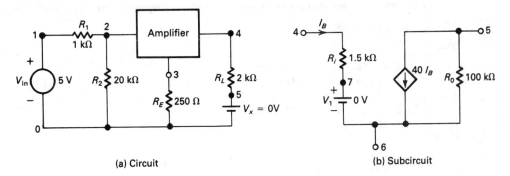

(a) Circuit   (b) Subcircuit

Figure 6.5 Circuit for Example 5.2

Solution The list of the statements for the circuit file is as follows.

```
*    Example 6.2    DC Sensitivity Analysis
*    Operating temperature is 40°.
.TEMP   40
*    Options
.OPTIONS  NOPAGE  NOECHO
*    DC input voltage of 5 V
VIN 1   0   DC   5V
R1   1   2   1K
R2   2   0   20K
RE   3   0   250
RL   4   5   2K
*    A dummy voltage source to measure the load current
VX   5   0   DC   0V
*    Subcircuit call for subcircuit CKT
X1   2   3   4   CKT
*    Subcircuit definition
.SUBCKT   CKT   4   6   5
RI   4   7   1.5K
RO   5   6   100K
*    Dummy voltage source for measuring controlling current
V1   7   6   DC   0V
*    I-controlled I-source
F1 5   6   V1   40
*    End of subcircuit definition
.ENDS   CKT
*    Sensitivity analysis calculates the DC bias point and prints
*    the current through the input source, I(VIN), before computing
*    the sensitivity.
*    It calculates and prints the sensitivity analysis of output
*    voltage V(4) with respect to all elements in the circuit.
.SENS   V(4)
.END
```

Note. The results of the sensitivity analysis are shown next. The node voltages are also printed automatically.

```
****     SMALL SIGNAL BIAS SOLUTION          TEMPERATURE =   27.000 DEG C
NODE    VOLTAGE      NODE   VOLTAGE      NODE    VOLTAGE      NODE    VOLTAGE
(  1)    5.0000   (    2)    4.3986   (    3)    3.8263   (    4)  -29.8470
(  5)    0.0000   (  X1.7)   3.8263
      VOLTAGE SOURCE CURRENTS
      NAME         CURRENT
      VIN         -6.014E-04
      VX          -1.492E-02
      X1.V1        3.815E-04
      TOTAL POWER DISSIPATION   3.01E-03  WATTS

****      DC SENSITIVITY ANALYSIS             TEMPERATURE =   40.000 DEG C
DC SENSITIVITIES OF OUTPUT V(4)
          ELEMENT          ELEMENT          ELEMENT          NORMALIZED
          NAME             VALUE            SENSITIVITY      SENSITIVITY
                                            (VOLTS/UNIT)     (VOLTS/PERCENT)
          R1               1.000E+03        3.590E-03        3.590E-02
          R2               2.000E+04       -6.564E-05       -1.313E-02
          RE               2.500E+02        9.600E-02        2.400E-01
          RL               2.000E+03       -1.486E-02       -2.972E-01
          X1.RI            1.500E+03        2.391E-03        3.587E-02
          X1.RS            1.000E+05       -1.426E-06       -1.426E-03
          VIN              5.000E+00       -5.969E+00       -2.985E-01
          VX               0.000E+00        9.958E-01        0.000E+00
          X1.V1            0.000E+00        6.268E+00        0.000E+00
      JOB CONCLUDED
      TOTAL JOB TIME              4.07
```

6.6.4 .TF Small-Signal Transfer Function

The small-signal transfer function capability of *PSpice* can be used to compute the small-signal DC gain, the input resistance, and the output resistance of a circuit. If V(1) and V(4) are the input and output variables, respectively, *PSpice* will calculate the small-signal DC gain between nodes 1 and 4, defined by

$$A_v = \frac{\Delta V_0}{\Delta V_i} = \frac{V(4)}{V(1)}$$

as well as the input resistance between nodes 1 and 0 and the small-signal dc output resistance between nodes 4 and 0.

PSpice calculates the small-signal DC transfer function by linearizing the circuit around the operating point. The statement for the transfer function has one of the forms

```
.TF   VOUT    VIN
.TF   IOUT    IIN
```

where VIN is the input voltage. VOUT (or IOUT) is the output voltage (or current). If the output is a current, then that current must be through a voltage source. The output variable, VOUT or IOUT, has the same format and meaning as in a .PRINT statement. If there are inductors and capacitors in a circuit, the inductors are treated as short circuits and capacitors as open circuits.

The .TF command calculates the parameters of the Thevenin's (or Norton's) equivalent circuit for the circuit file. It prints automatically the output and does not require .PRINT or .PLOT or .PROBE statements.

Statements for Transfer Function Analysis

```
.TF   V(10)    VIN
.TF   I(VX)    IIN
```

Example 6.3

For the circuit in Figure 6.5, calculate and print (a) the voltage gain, $A_v = V(4)/V_{in}$; (b) the input resistance, R_{in}; and (c) the output resistance, R_o.

Solution   The list of the circuit file is as follows.

```
*    Example 6.3  Transfer Function Analysis
.OPTIONS   NOPAGE   NOECHO
*    DC input voltage of 5 V
VIN 1   0   DC   5V
R1   1   2   1K
R2   2   0   20K
RE   3   0   250
RL   4   5   2K
*    A dummy voltage source to measure the load current
VX   5   0   DC   0V
*    Subcircuit call for subcircuit CKT
X1   2   3   4   CKT
*    Subcircuit definition
.SUBCKT   CKT   4   6   5
RI   4   7   1.5K
RO   5   6   100K
*    Dummy voltage source for measuring controlling current
V1   7   6   DC   0V
*    I-controlled I-source
F1   5   6   V1   40
*    End of subcircuit definition
.ENDS   CKT
*    The .TF command calculates and prints the dc gain, the input
```

```
*    resistance, and the output resistance.  The input voltage
*    is VIN and output voltage is V(4).
.TF  V(4)   VIN
.END
```

Note. The results of the .TF command are shown next.

```
****      SMALL-SIGNAL CHARACTERISTICS
          V(4)VIN = -5.969E+00
          INPUT RESISTANCE AT VIN =  8.313E+03
          OUTPUT RESISTANCE AT V(4) =  1.992E+03
             JOB CONCLUDED
             TOTAL JOB TIME              4.01
```

6.6.5 .DC DC Sweep

DC sweep is also known as the *DC transfer characteristic*. The input variable is varied over a range of values. For each value of input variables, the DC operating point and the small-signal DC gain are computed by calling the small-signal transfer function capability of *PSpice*. The DC sweep (or DC transfer characteristic) is obtained by repeating the calculations of small-signal transfer function for a set of values. The statement for performing DC sweep takes one of the general forms

```
.DC  VIN  VSTART  VSTOP  VINCR
.DC  IIN  ISTART  ISTOP  IINCR
.DC  VIN  VSTART  VSTOP  VINCR  IIN  ISTART  ISTOP  IINCR
.DC  IIN  ISTART  ISTOP  IINCR  VIN  VSTART  VSTOP  VINCR
```

This command allows one to vary an independent voltage source VIN from VSTART to VSTOP with an increment of VINCR and an independent current source, IIN, from ISTART to ISTOP with an increment of IINCR.

VSTART (or ISTART) may be greater than or less than VSTOP (or ISTOP). That is, the sweep may go into either direction. However, the increment VINCR (or IINCR) must be positive only and must not be zero or negative.

The DC sweep can be nested, similar to a DO loop within a DO loop in FORTRAN programming. The first sweep is the inner loop and the second sweep is the outer loop. The first sweep is done for each value of the second sweep.

PSpice does not print or plot any output by itself for DC sweep. The results of DC sweep are obtained by .PRINT, .PLOT, or .PROBE statements.

Statements for DC Sweep

```
.DC  VIN  -5V    10V    0.25V
.DC  IIN  50MA  -50MA   1MA
.DC  VA   0  15V  0.5V  IA  0  1MA  0.05MA
.DC  V2   10V  12V  5V
```

Notes

1. If the source has a DC value, its value is set by the sweep overriding the DC value.
2. In the third statement, the current source IA is the inner loop and the voltage source VA is the outer loop. *PSpice* will vary the value of the current source IA from 0 to 1 MA with an increment of 0.05 MA for each value of voltage source VA, and generate an entire print table or plot for each value of voltage sweep.

Example 6.4

For the circuit in Figure 6.5, calculate and plot the DC transfer characteristic V_o versus V_{in}. The input voltage is varied from 0 to 10 V with an increment of 0.5 V. **Solution**   The list of the circuit file is as follows.

```
*     Example 6.4    DC Sweep
*     The analysis is performed for three operating temperatures
.OPTIONS  NOPAGE   NOECHO
*     DC input voltage of 5 V
VIN 1   0   DC   5V
R1   1   2   1K
R2   2   0   20K
RE   3   0   250
RL   4   5   2K
*     A dummy voltage source to measure the load current
VX   5   0   DC   0V
*     Subcircuit call for subcircuit CKT
X1   2   3   4   CKT
*     Subcircuit definition
.SUBCKT   CKT   4   6   5
RI   4   7   1.5K
RO   5   6   100K
*  Dummy voltage source for measuring controlling current
V1   7   6   DC   0V
*  I-controlled I-source
F1 5   6   V1   40
*  End of subcircuit definition
.ENDS  CKT
*  DC sweep from 0 to 10 V with an increment of 0.5 V
.DC   VIN   0   10V   0.5V
*  PSpice plots the results of dc sweep
.PLOT   DC   V(4)
.PROBE
.END
```

The transfer characteristic is shown in Figure 6.6. The details of the bias points can be printed on the output file by .OP command.

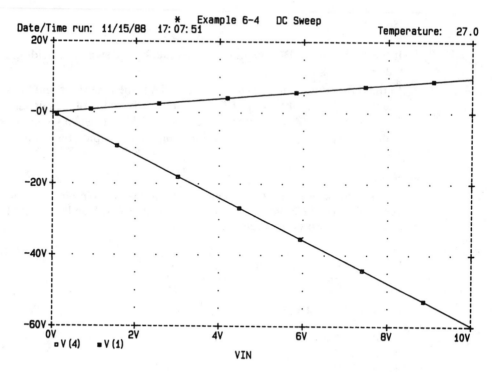

Figure 6.6   DC transfer characteristic for Example 6.4

6.7 AC ANALYSIS

The AC analysis calculates the frequency response of a circuit over a range of frequencies. If the circuit contains nonlinear devices or elements, it is necessary to obtain the small-signal parameters of the elements before calculating the frequency response. Prior to the frequency response (or AC analysis), *PSpice* determines the small-signal parameters of the elements. The method for calculation of bias point for AC analysis is identical to that for DC analysis. The details of the bias points can be printed by the .OP command.

The command for performing frequency response takes one of the general forms

```
.AC   LIN   NP   FSTART   FSTOP
.AC   OCT   NP   FSTART   FSTOP
.AC   DEC   NP   FSTART   FSTOP
```

NP is the number of points in a frequency sweep. FSTART is the starting frequency and FSTOP is the ending frequency. Only one of LIN, OCT, or DEC must be specified in the statement. LIN, OCT, or DEC specify the type of sweep as follows:

>   LIN **Linear sweep:** The frequency is swept linearly from the starting frequency to the ending frequency and NP becomes the total number of points in the sweep. The next frequency is generated by adding a constant to the present value. LIN is used if the frequency range is narrow.

>   OCT **Sweep by octave:** The frequency is swept logarithmically by octave, and NP becomes the number of points per octave. The next frequency is generated by multiplying the present value by a constant larger than unity. OCT is used if the frequency range is wide.

>   DEC **Sweep by decade:** The frequency is swept logarithmically by decade, and NP becomes the number of points per decade. DEC is used if the frequency range is the widest.

PSpice does not print or plot any output by itself for AC analyses. The results of an AC sweep are obtained by .PRINT, .PLOT, or .PROBE statements.

Some Statements for AC Analysis

```
.AC   LIN   201  100HZ   300HZ
.AC   LIN   1    60HZ    120HZ
.AC   OCT   10   100HZ   10KHZ
.AC   DEC   100  1KHZ    1MEGHZ
```

Notes

1. FSTART must be less than FSTOP and can not be zero.
2. NP = 1 is permissible, and the second statement calculates the frequency response at 60 Hz only.
3. Before performing the frequency response analysis, *PSpice* automatically calculates the biasing point to determine the linearized circuit parameters around the bias point.
4. All independent voltage and current sources that have AC values are inputs to the circuit. At least one source must have an AC value; otherwise, the analysis would not be meaningful.
5. If a group delay output is required by a suffix of G, as mentioned in Section 3.3, the frequency steps should be small so that the output changes smoothly.

6.8 NOISE ANALYSIS

Resistors and semiconductor devices generate noise. The level of the noise depends on the frequency. The various types of noise that are generated by resistors and semiconductor devices are discussed in Appendix B. Noise analysis is done in conjunction with AC analysis and requires an .AC command. For each frequency of the AC analysis, the noise level of each generator in a circuit (e.g., resistors and transistors) is calculated, and their contributions to the output nodes are computed by summing the RMS noise values. The gain from the input source to the output voltage is calculated. From this gain, the equivalent input noise level at the specified source is calculated by *PSpice*.

The statement for performing noise analysis is of the form

```
.NOISE   V(N+, N-)   SOURCE    M
```

where V(N+, N-) is the output voltage across nodes N+ and N-. The output could be at a node N, such as V(N).

SOURCE is the name of an independent voltage or current source at which the equivalent input noise will be generated. It should be noted that SOURCE is not a noise generator; rather it is a place at which to compute the equivalent noise input. For a voltage source, the equivalent input is in $V/\sqrt{Hz}$; for a current source, it is in $A/\sqrt{Hz}$.

M is the print interval that permits to print a table for the individual contribution of all generators to the output nodes for every *m*th frequency. There is no need for .PRINT or .PLOT command for printing a table of all contributions. If the value of M is not specified, then *PSpice* does not print a table of individual contribution. The output noise and equivalent noise can also be printed by .PRINT or .PLOT command.

Statements for Noise Analysis

```
.NOISE   V(4,5)   VIN
.NOISE   V(6)     IIN
.NOISE   V(10)    V1       10
```

Note. The .PROBE command cannot be used for noise analysis.

Example 6.5

For the circuit in Figure 6.5, calculate and print the equivalent input and output noise if the frequency of the source is varied from 1 Hz to 100 kHz. The frequency should be increased by a decade with 1 point per decade.

Solution   The input source is of AC type. The list of the circuit file is as follows.

```
*   Example 6.5   Noise Analysis
.OPTIONS  NOPAGE  NOECHO
*   The input voltage VIN, which is connected between nodes 1 and 0,
*   has a value of 5 V for DC analysis, a value of 1 V for AC analysis.
VIN 1   0   DC  5V
R1  1   2   1K
R2  2   0   20K
RE  3   0   250
RL  4   0   2K
```

```
*    Subcircuit call for subcircuit CKT
X1   2   3   4   CKT
*    Subcircuit definition
.SUBCKT   CKT   4   6   5
RI   4   7   1.5K
RO   5   6   100K
*  Dummy voltage source for measuring controlling current
V1   7   6   DC   0V
*  I-controlled I-source
F1   5   6   V1   40
*  End of subcircuit definition
.ENDS   CKT
*  AC sweep from 1 Hz to 100 kHz with a decade increment and 11 points
*  per decade
.AC   DEC   1   1HZ   100KHz
*  Noise analysis without printing details of individual contributions
.NOISE   V(4)   VIN
*  PSpice prints the details of equivalent input and output noise
.PRINT   NOISE   ONOISE   INOISE
.END
```

The results of the noise analysis are shown next. The node voltages are also printed automatically by *PSpice*.

```
****      AC ANALYSIS                    TEMPERATURE =    27.000 DEG C
   FREQ          ONOISE         INOISE
   1.000E+00     4.234E-08      7.093E-09
   1.000E+01     4.234E-08      7.093E-09
   1.000E+02     4.234E-08      7.093E-09
   1.000E+03     4.234E-08      7.093E-09
   1.000E+04     4.234E-08      7.093E-09
   1.000E+05     4.234E-08      7.093E-09
             JOB CONCLUDED
             TOTAL JOB TIME          5.11
```

6.9 TRANSIENT RESPONSE

A transient response determines the output in time domain in response to an input signal in time domain. The method for the calculation of transient analysis bias point differs from that of DC analysis bias point. The DC bias point is also known as the **regular bias point.** In the regular (DC) bias point, the initial values of the circuit nodes do not contribute to the operating point and to the linearized parameters. The capacitors and inductors are considered open-circuited and short-circuited, respectively, whereas in the transient bias point, the initial values of the circuit nodes are taken into account in calculating the bias point and the small-signal parameters of the nonlinear elements. The capacitors and inductors that may have initial values therefore remain as parts of the circuit.

The determination of the transient analysis requires statements involving

.IC   Initial transient conditions
.TRAN   Transient analysis

6.9.1 .IC Initial Transient Conditions

The various nodes can be assigned to initial voltages during the transient analysis, and the general form for assigning initial values is

```
.IC   V(1)=V1   V(1)=V2  ···  V(N)=VN
```

where V1, V2, V3, . . . are the initial voltages for nodes V(1), V(2), V(3), . . . , respectively. These initial values are used by *PSpice* to calculate the transient analysis bias point and the linearized parameters of nonlinear devices for transient analysis. After the transient analysis bias point has been calculated, the transient analysis starts and the nodes are released. It should be noted that these initial conditions do not affect the regular bias point calculation during DC analysis or DC sweep. For the .IC statement to be effective, UIC (use initial conditions) *should not* be specified in the .TRAN command.

Statement for Initial Transient Conditions

```
.IC   V(1)=2.5   V(5)=1.7V   V(7)=0.5
```

6.9.2 .TRAN Transient Analysis

The transient analysis can be performed by the .TRAN command that has one of the general forms

```
.TRAN       TSTEP   TSTOP  [TSTART TMAX]  [UIC}
.TRAN[/OP]  TSTEP   TSTOP  [TSTART TMAX]  [UIC}
```

TSTEP is the printing increment, and TSTOP is the final time (or stop time). TMAX is the maximum size of internal time step. TMAX allows the user to control the internal time step. TMAX can be smaller or larger than the printing time, TSTEP. The default value of TMAX is TSTOP/50.

The transient analysis always starts at time = 0. However, it is possible to suppress the printing of the output for a time of TSTART. TSTART is the initial time at which the transient response is printed. In fact *PSpice* analyzes the circuit from t = 0 to TSTART, but it does not print or store the output variables. Although *PSpice* computes the results with an internal time step, the results are generated by interpolation for a printing step of TSTEP. Figure 6.7 shows the relationships of TSTART, TSTOP, and TSTEP.

In transient analysis, only the node voltages of the transient analysis bias point are printed. However, the .TRAN command can control the output for the transient response bias point. An .OP command with .TRAN command, namely, .TRAN/OP, will print the small-signal parameters during transient analysis.

If UIC is not specified as an optional at the end of .TRAN statement, *PSpice* calculates the transient analysis bias point before the beginning of transient analysis. *PSpice* uses the initial values specified with the .IC command.

If UIC (use initial conditions) is specified as an option at the end of .TRAN statement, *PSpice* does not calculate the transient analysis bias point before the beginning of transient analysis. However, *PSpice* uses the initial values specified

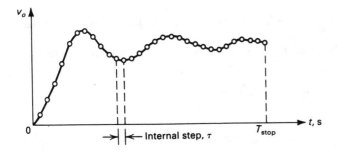

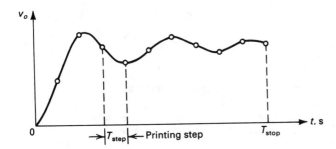

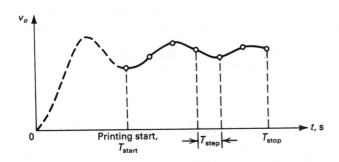

Figure 6.7 Response of transient analysis

with the IC= initial conditions for capacitors and inductors, which are discussed in Chapter 5. Therefore, if UIC is specified, the initial values of the capacitors and inductors *must* be supplied. The .TRAN statement requires .PRINT or .PLOT or .PROBE statement to get the results of the transient analysis.

Statements for Transient Analysis

```
.TRAN     5US     1MS
.TRAN     5US     1MS  200US  0.1NS
.TRAN     5US     1MS  200US  0.1NS  UIC
.TRAN/OP  5US     1MS  200US  0.1NS  UIC
```

Example 6.6

Repeat Example 5.1 if the voltage across the capacitor is set by .IC command instead of IC condition and UIC is not specified.
Solution   The list of the circuit file with .IC statement and without UIC is as follows.

Example 6.6   Transient Response of RLC Circuit

```
*   The operating temperature is 50°.
.TEMP  50
*   Input step voltage represented by an PWL waveform
VS   1   0   PWL (0   0   10NS   10V   2MS  10V)
R1   1   2   RMOD  6OHM
*   Inductor of 1.5 mH with an initial current of 3 A
L1   2   3   LMOD  1.5MH   IC=3A
*   Capacitor of 2.5 µF with an initial voltage of 4 V
C1   3   0   CMOD  2.5UF   IC=4V
R2   3   0   RMOD  2OHM
*   Model statements for resistor, inductor, and capacitor
.MODEL   RMOD   RES (R=1   TC1=0.02   TC2=0.005)
.MODEL   CMOD   CAP (C=1   VC1=0.01   VC2=0.002   TC1=0.02   TC2=0.005)
.MODEL   LMOD   IND (L=1   IL1=0.1    IL2=0.002   TC1=0.02   TC2=0.005)
*   The initial voltage at node 3 is 4 V.
.IC   V(3)=4V
*   Transient analysis from 0 to 2 ms with 10 = µs time increment and
*   without using initial conditions (UIC).  That is, IC=4V has no effect.
.TRAN  5US  1MS
*   Plot the results of transient analysis - voltage at node 3.
.PLOT   TRAN   V(3)   V(1)
.PROBE
.END
```

The results of the simulation are shown in Figure 6.8. It can be noticed that the response is completely different from that of Fig. 5-3.

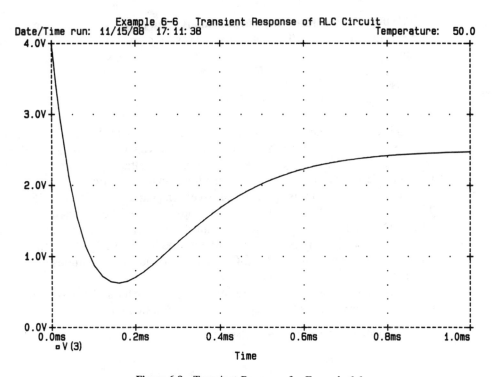

Figure 6.8 Transient Response for Example 6.6

6.10 FOURIER ANALYSIS

The output variables from the transient analysis are in discrete forms. These sampled data can be used to calculate the coefficients of Fourier series. A periodic waveform can be expressed in a Fourier series as

$$v(\theta) = C_0 + \sum_{n=1}^{\infty} C_n \sin(n\theta + \phi_n)$$

where

$$\theta = 2\pi ft$$

$$f = \text{frequency in hertz}$$

$$C_o = \text{DC component}$$

$$C_n = n\text{th harmonic component}$$

PSpice uses the results of the transient analysis to perform the Fourier analysis up to ninth harmonics or 10 coefficients. The statement takes one of these general forms:

```
.FOUR   FREQ   V1   V2   V3···VN
.FOUR   FREQ   I1   I2   I3···IN
```

FREQ is the fundamental frequency. V1, V2, . . . (or I1, I2, . . .) are the output voltages (or currents) for which the Fourier analysis is desired. A .FOUR statement must have a .TRAN statement. The output voltages (or currents) must have the same forms as in .TRAN statement for transient analysis.

Fourier analysis is performed over the interval (TSTOP-PERIOD) to TSTOP, where TSTOP is the final (or stop) time for the transient analysis and PERIOD is one period of the fundamental frequency. Therefore, the duration of the transient analysis must be at least one period long, PERIOD. At the end of the analysis, *PSpice* determines the DC component and the amplitudes of up to ninth harmonics.

PSpice does print a table, automatically, the results of Fourier analysis and does not require .PRINT, .PLOT, or .PROBE statement.

Statement for Fourier Analysis

```
.FOUR   100KHZ   V(2,3),   V(3),   I(R1),   I(VIN)
```

Example 6.7

For Example 5.3, calculate the coefficients of the Fourier series if the fundamental frequency is 1 kHz.
Solution   The list of the circuit file is as follows.

```
      Example 6.7     Fourier Analysis
*   Sinusoidal input voltage of 200-V peak with 0 phase delay
VS   1   0   SIN (0   200V   1KHZ)
RS   1   2   100OHM
R1   2   0   100KOHM
*   Voltage-controlled source with a voltage gain of 0.1
E1   3   0   2   0   0.1
RL   4   5   2OHM
*   A dummy voltage source to measure the load current
VX   5   0   DC   0V
*   Voltage-controlled switch controlled by voltage across nodes 3 and 0
S1   3   4   3   0   SMOD
*   Switch model descriptions
.MODEL   SMOD   VSWITCH (RON=5M ROFF=10E+9 VON=25M  VOFF=0.0)
*   Transient analysis from 0 to 16 ms with 0.1-ms increment
.TRAN   5US   1MS
*   Fourier analysis of load current at a fundamental frequency of 1 kHz
.FOUR   1KHZ   I(VX)
.END
```

The results of Fourier analysis are shown next.

```
****      FOURIER ANALYSIS              TEMPERATURE =    27.000 DEG C
FOURIER COMPONENTS OF TRANSIENT RESPONSE I(VX)
DC COMPONENT =    3.171401E+00
HARMONIC   FREQUENCY      FOURIER    NORMALIZED      PHASE        NORMALIZED
   NO        (HZ)       COMPONENT    COMPONENT      (DEG)       PHASE (DEG)
    1     1.000E+03     4.982E+00    1.000E+00     2.615E-05     0.000E+00
    2     2.000E+03     2.115E+00    4.245E-01    -9.000E+01    -9.000E+01
    3     3.000E+03     9.002E-08    1.807E-08     6.216E+01     6.216E+01
    4     4.000E+03     4.234E-01    8.499E-02    -9.000E+01    -9.000E+01
    5     5.000E+03     9.913E-08    1.990E-08     8.994E+01     8.994E+01
    6     6.000E+03     1.818E-01    3.648E-02    -9.000E+01    -9.000E+01
    7     7.000E+03     7.705E-08    1.547E-08     5.890E+01     5.890E+01
    8     8.000E+03     1.012E-01    2.032E-02    -9.000E+01    -9.000E+01
    9     9.000E+03     9.166E-08    1.840E-08     7.348E+01     7.348E+01
    TOTAL HARMONIC DISTORTION =    4.349506E+01 PERCENT
         JOB CONCLUDED
         TOTAL JOB TIME          45.54
```

SUMMARY

.AC	AC analysis
.DC	DC analysis
.END	End of circuit
.ENDS	End of subcircuit
.FOUR	Fourier analysis
.IC	Initial transient conditions
.INC	Include file
.LIB	Library file
.MODEL	Model
.NODESET	Nodeset
.NOISE	Noise analysis
.OP	Operating point
.OPTIONS	Options
.PLOT	Plot
.PRINT	Print
.PROBE	*Probe*
.SENS	Sensitivity analysis
.SUBCKT	Subcircuit definition
.TEMP	Temperature
.TF	Transfer function
.TRAN	Transient analysis
.WIDTH	Width

PROBLEMS

6.1. For the circuit in Figure P6.1, calculate and print the sensitivity of output voltage V_o with respect to each circuit element. The operating temperature is 50°C.

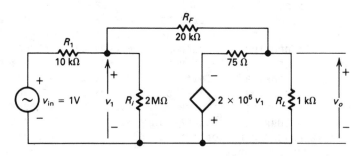

Figure P6.1

6.2. For the circuit in Figure P6.1, calculate and print (a) the voltage gain, $A_v = V_o/V_{in}$; (b) the input resistance, R_{in}; and (c) the output resistance, R_o.

6.3. For the circuit in Figure P6.1, calculate and plot the DC transfer characteristic V_o versus V_{in}. The input voltage is varied from 0 to 10 V with an increment of 0.5 V.

6.4. For the circuit in Figure P6.1, calculate and print the equivalent input and output noise if the frequency of the source is varied from 10 Hz to 1 MHz. The frequency should be increased by a decade with 2 points per decade.

6.5. For the circuit in Figure P6.5, the frequency response is to be calculated and printed over the frequency range from 1 Hz to 100 kHz with a decade increment and 10 points per decade. The peak magnitude and phase angle of the output voltage is to be plotted on the output file. The results should also be available for display and hard copy by .PROBE command.

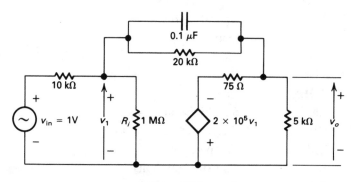

Figure P6.5

6.6. Repeat Problem 6.5 for the circuit in Figure P6.6.

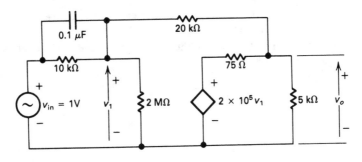

Figure P6.6

6.7. For the circuit in Figure P6.5, calculate and plot the transient response of the output voltage from 0 to 2 ms with a time increment of 5 μs. The input voltage is shown in Figure P6.7. The results should be available for display and hard copy by *Probe*.

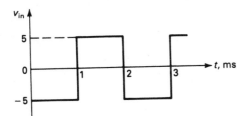

Figure P6.7

6.8. Repeat Problem 6.7 for the input voltage as shown in Figure P6.8.

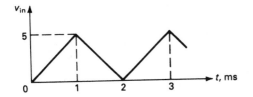

Figure P6.8

6.9. For the circuit in Figure P6.6, calculate and plot the transient response of the output voltage from 0 to 2 ms with a time increment of 5 μs. The input voltage is shown in Figure P6.7. The results should be available for display and hard copy by *Probe*.

6.10. Repeat Problem 6.9 for the input voltage as shown in Figure P6.8.

6.11. For Problem 6.7, calculate the coefficients of the Fourier series if the fundamental frequency is 500 Hz.

6.12. For Problem 6.8, calculate the coefficients of the Fourier series if the fundamental frequency is 500 Hz.

6.13. For Problem 6.9, calculate the coefficients of the Fourier series if the fundamental frequency is 500 Hz.

6.14. For Problem 6.10, calculate the coefficients of the Fourier series if the fundamental frequency is 500 Hz.

7

Semiconductor Diodes

7.1 INTRODUCTION

A semiconductor diode may be specified in *PSpice* by a diode statement in conjunction with a model statement. The diode statement specifies the diode name, the nodes to which the diode is connected, and its model name. The model incorporates an extensive range of diode characteristics, e.g., DC and small-signal behavior, temperature dependency, and noise generation. The model parameters take into account temperature effects, various capacitances, and physical properties of semiconductors.

7.2 DIODE MODEL

The PSpice model for a reversely biased diode is shown in Figure 7.1 [1, 2]. The small-signal model and the static model, which are generated by *PSpice,* are shown in Figures 7.2 and 7.3, respectively. In the static model, the diode current that depends on its voltage is represented by a current source. The small-signal parameters are generated by *PSpice* from the operating point.

PSpice generates a complex model for diodes. The model equations that are used by *PSpice* are described in [1] and [2]. In many cases, especially at the level

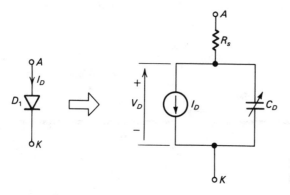

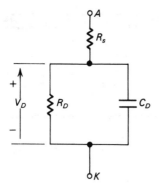

Figure 7.1 PSpice diode model with reverse-biased diode

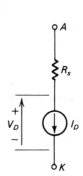

Figure 7.2 Small-signal diode model

Figure 7.3 Static diode model with reverse-biased diode

at which this book is aimed, such complex models are not necessary. Many model parameters can be ignored by the users, and *PSpice* assigns default values of the parameters.

The model statement of a diode has the general form

```
.MODEL DNAME  D (P1=V1 P2=V2 P3=V3 ........PN=VN)
```

where DNAME is the model name. DNAME can begin with any character, but its word size is normally limited to 8. D is the type symbol for diodes. P_1, P_2, . . . and V_1, V_2, . . . are the model parameters and their values, respectively. The model parameters are listed in Table 7.1.

An **area factor** is used to determine the number of equivalent parallel diodes of a specified model. The model parameters that are affected by the area factor are marked by an asterisk (*) in the descriptions of the model parameters.

The diode is modeled as an ohmic resistance (value = RS/area) in series with an intrinsic diode. The resistance is attached between node NA and an internal

TABLE 7.1 PARAMETERS OF DIODE MODEL

Name	Area	Model parameter	Units	Default	Typical
IS	*	Saturation current	Amps	1E-14	1E-14
RS	*	Parasitic resistance	Ohms	0	10
N		Emission coefficient		1	
TT		Transit time	seconds	0	0.1NS
CJO	*	Zero-bias *pn* capacitance	Farads	0	2PF
VJ		Junction potential	Volts	1	0.6
M		Junction grading coefficient		0.5	0.5
EG		Activation energy	Electron-volts	1.11	11.1
XTI		IS temperature exponent		3	3
KF		Flicker noise coefficient		0	
AF		Flicker noise exponent		1	
FC		Forward bias depletion capacitance coefficient		0.5	
BV		Reverse breakdown voltage	Volts	∞	50
IBV	*	Reverse breakdown current	Amps	1E-10	

anode node. [(area) value] scales IS, RS, CJO, and IBV, and defaults to 1. IBV and BV are both specified as positive values.

The DC characteristic of a diode is determined by the reverse saturation current IS, the emission coefficient N, and the ohmic resistance RS. The charge storage effects are modeled by the transit time TT, a nonlinear depletion layer capacitance, which depends on the zero-bias junction capacitance CJO, the junction potential VJ, and grading coefficient M. The temperature of the reverse saturation current is defined by the gap activation energy (or gap energy) EG and saturation temperature exponent XTI.

In order to simulate a Zener diode, the model in Figure 7.4 can be used. Diode D_1 and the threshold voltage source V_o represent the normal forward voltage and reverse behavior of Zener diode. Diode D_2, the voltage source BV, and

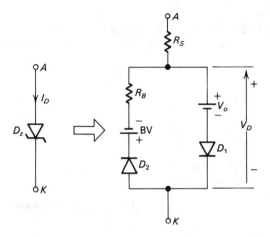

Figure 7.4 Static Zener diode model

resistance R_B define the breakdown region. Diode D_2 does not conduct until $V_D =$ $-$BV, and if the reverse voltage is increased, then diode D_2 becomes forward biased and the reverse current flows through R_B.

Reverse breakdown is modeled by an exponential increase in the reverse diode current and is determined by the reverse breakdown voltage, BV, and the current at breakdown voltage, IBV.

7.3 DIODE STATEMENT

The symbol for a diode is D. The name of a diode must start with D and it takes the general form

```
D<name> NA  NK  DNAME  [(area) value]
```

where NA and NK are the anode and cathode nodes, respectively. The current flows from anode node NA through the diode to cathode node NK. DNAME is the model name.

Some Statements for Diode

```
DM    5    6    DNAME
.MODEL DNAME  D (IS=0.5UA RS=6 BV=5.20 IBV=0.5UA)
D15   33   35  SWITCH  1.5
.MODEL SWITCH D (IS=100E-15 RS=16 CJO=2PF TT=12NS BV=100 IBV=100E-15)
DCLAMP  0   8   D1N914
.MODEL D1N914 D (IS=100E-15 RS=16 CJO=2PF TT=12NS BV=100 IBV=100E-15)
```

Note. Diode DM, having model name DNAME, is a Zener diode with a zener breakdown voltage of 5.2 V; the current at the zener break is 0.5 μA.

Example 7.1

A diode circuit is shown in Figure 7.5. Plot the V-I characteristic of the diode for forward voltage from 0 to 2 V and for temperature of 50, 100, and 150°. The diode is of type D1N914 and the model parameters are IS=100E$-$15 RS=16 BV=100 IBV=100E$-$15.

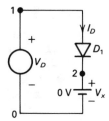

Figure 7.5 Diode circuit

Solution   The list of the circuit file is as follows.

Example 7.1   Diode Characteristic
```
.OPTIONS NOPAGE  NOECHO
*   Operating temperatures: 50, 100, and 150°
.TEMP  50  100 150
*   The input voltage that will be overridden by DC sweep voltage
*   is assumed zero.
VD  1  0  DC  0V
*   Diode D1 whose model name is D1N914 is connected between nodes 1 and 2.
D1   1  2  D1N914
*   A dummy voltage source to measure the diode current
VX  2  0  DC  0V
*   DIode model defines the model parameters.
.MODEL D1N914 D (IS=100E-15 RS=16 BV=100 IBV=100E-15)
*   DC sweep from 0 to 2 V with 0.01 V increment
.DC  VD  0  2V  0.01V
*   Plot the diode current from the results of DC sweep.
.PLOT  DC  I(VX)
*   Graphic post-processor
.PROBE
.END
```

The V-I characteristic that is obtained by varying the input voltage is shown in Figure 7.6.

Example 7.2

A Zener voltage regulator is shown in Figure 7.7. Plot the DC transfer characteristic if the input voltage is varied from -15 V to 15 V with an increment of 0.5 V. The Zener voltages of the diodes are the same and $V_Z = 5.2$ V; the current at the zener breakdown is $I_Z = 0.5$ μA. The model parameters are IS=0.5UA RS=6 BV=5.20 IBV=0.5UA. The operating temperature is 50°C.

Solution   A Zener diode is implemented by setting the model parameters BV = V_Z = 5.2 V and IBV = I_Z = 0.5 μA. The list of the circuit file is as follows.

Example 7.2   Zener Regulator
```
.OPTIONS NOPAGE  NOECHO
*   Operating temperatures: 50°
.TEMP  50
VIN  1  0  DC 0V
R1   1  2  500
*   Diodes D1 and D2 have model name DNAME.
D1   2  3  DNAME
D2   0  3  DNAME
RL   2  0  1K
```

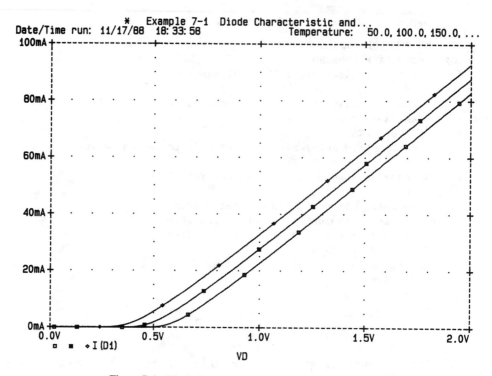

Figure 7.6   Diode forward characteristic for Example 7.1

```
*    Model DNAME defines the parameters of Zener diodes.
.MODEL DNAME  D (IS=0.5UA RS=6 BV=5.20 IBV=0.5UA)
*    DC sweep from -15 to 15 V with 0.5-V increment
.DC  VIN  -15  15V  0.5V
*    Print the load voltage using the results of DC sweep.
.PRINT  DC  V(2)
.PROBE
.END
```

The DC transfer characteristic is shown in Figure 7.8.

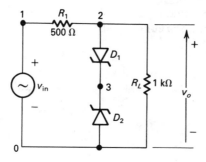

Figure 7.7   Zener regulator

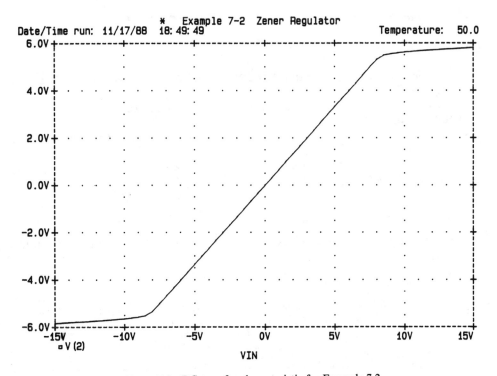

Figure 7.8 DC transfer characteristic for Example 7.2

Example 7.3

A clamping circuit is shown in Figure 7.9, where the output is taken from node 2. Plot the transient response of the output voltage V(2) for the time duration of 0 to 3 ms in steps of 20 μs. The initial capacitor voltage is -15 V. The model parameters of the diode are the default values.

Solution   The list of circuit file is as follows.

```
Example 7.3    Diode Clamper Circuit
*    Input voltage of 10 V peak at 1 kHz and zero offset voltage is
*    connected between nodes 1 and 0.
VIN  1  0  SIN (0 10 1KHZ)
```

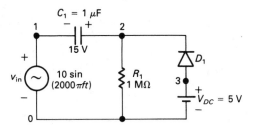

Figure 7.9 Diode clamper circuit

```
*    Capacitor with an initial voltage of -15 V
C1   1   2   1UF IC=-15V
R1   2   0   1MEG
VDC  3   0   DC 5
*    Diode with model name DIODE is connected between nodes 3 and 2.
D1   3   2   DIODE
*    Diode model with default values of parameters
.MODEL DIODE   D
*    Transient analysis for 0 to 3 ms with 20-µs increment with UIC
.TRAN  20US  3MS  UIC
*    Plot transient voltages at nodes 1 and 2.
.PLOT TRAN   V(2)   V(1)
.PROBE
.END
```

The input and output voltages of the diode clamper circuit are shown in Figure 7.10.

Example 7.4

A diode circuit is shown in Figure 7.11(a). The AC input voltage is $v_{in} = 10 \times 10^{-3}\sin(2\pi \times 10^3 t)$. Print the DC bias point and the small-signal parameters of the

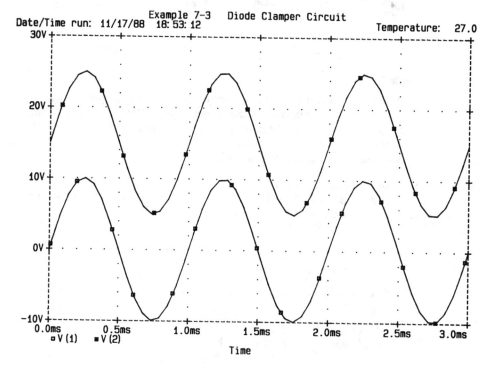

Figure 7.10   Output of diode clamper circuit for Example 7.3

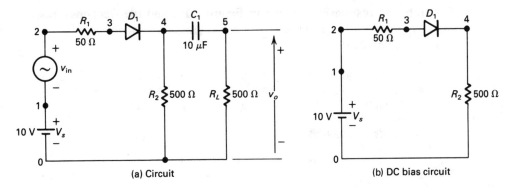

Figure 7.11   Diode circuit

diode. Plot the output voltage from 0 to 200 μs with 2-μs increments. If the frequency of the AC voltage is varied from 1 Hz to 1 kHz, plot the magnitude and phase angle of the output voltage. The model parameters are IS=100E−15 RS=16 CJO=2PF TT=12NS BV=100 IBV=100E−15.

Solution   The equivalent circuit for calculating the dc bias point and small-signal parameters is shown in Figure 7.11(b). The list of the circuit file is as follows.

Example 7.4   Diode Circuit
```
*   DC voltage of 10 V
VS   1   0   DC 10V
*   AC voltage of 10 mV peak at 10 kHz and zero offset voltage for
*   transient analysis and 10 mV peak for AC analysis
VIN   2   1   AC   10MV   SIN (0 10M 10KHZ)
R1   2   3   50
R2   4   0   500
C1   4   5   10UF
RL   5   0   500
*   Diode with model name D1N914 is connected between nodes 3 and 4.
D1   3   4   D1N914
*   DIode model defines the model parameters.
.MODEL D1N914 D (IS=100E-15 RS=16 CJO=2PF TT=12NS BV=100 IBV=100E-15)
*   Transient analysis for 0 to 200 μs with 2-μs increments.
.TRAN   2US   200US
*   AC analysis from 1 Hz to 1 kHz with 10 points per decade
.AC   DEC   10   1HZ   1kHZ
*   Plot transient voltages
.PLOT TRAN   V(5)   V(2,1)
*   Magnitude and phase plots of output voltage at node 5
.PLOT   AC   VDB(5)   VP(5)
*   Printing of the small-signal parameters and the dc operating point.
.OP
.PROBE
.END
```

The transient response is shown in Figure 7.12, and the frequency response is shown in Figure 7.13. The details of the DC bias point and the small-signal parameters are given next.

```
****      SMALL SIGNAL BIAS SOLUTION      TEMPERATURE =   27.000 DEG C
  NODE    VOLTAGE       NODE    VOLTAGE      NODE    VOLTAGE      NODE    VOLTAGE
(    1)   10.0000  (     2)   10.0000  (    3)     9.1756  (    4)     8.2438
(    5)    0.0000
     VOLTAGE SOURCE CURRENTS
     NAME          CURRENT
     VS            -1.649E-02
     VIN           -1.649E-02
     TOTAL POWER DISSIPATION    1.65E-01   WATTS
****      OPERATING POINT INFORMATION      TEMPERATURE =   27.000 DEG C
  NAME          D1
  MODEL         D1N914
  ID            1.65E-02
  VD            9.32E-01
  REQ           1.57E+00
  CAP           7.65E-09
```

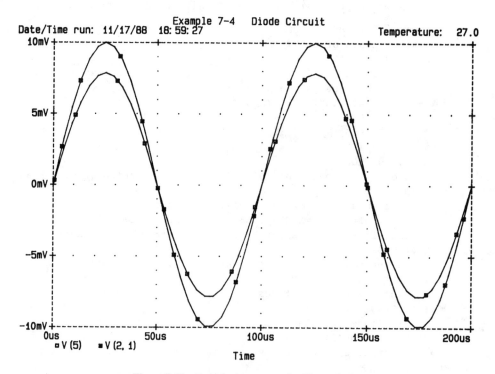

Example 7-4   Diode Circuit
Date/Time run: 11/17/88 18:59:27   Temperature: 27.0

Figure 7.12 Transient response for Example 7.4

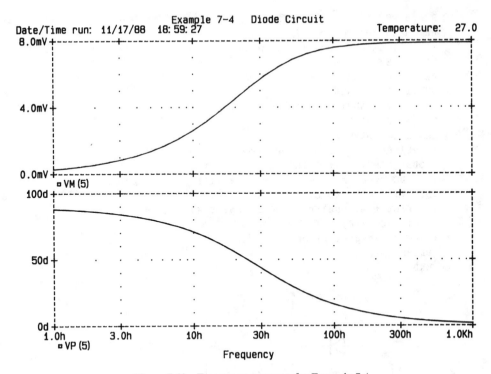

Figure 7.13   Frequency response for Example 7.4

Example 7.5

A diode waveform-shaping circuit is shown in Figure 7.14, where the output is taken from node 2. Plot the transfer characteristic between V_{in} and V(2) for values of V_{in} in the range of −15 V to 30 V in steps of 0.5 V. The model parameters of the diodes are IS=100E−15 RS=16 BV=100 IBV=100E−15.
Solution   The list of circuit file is as follows.

```
Example 7.5    Diode Waveform-Shaping Circuit
*    The input voltage that is overridden by DC sweep voltage is assumed 0 V.
VIN   1   0   DC   0
VCC   6   0   DC   15V
VEE   10  0   DC   -15V
R1    6   5   2K
R2    5   4   2K
R3    4   3   2K
R4    3   2   2K
R5    2   7   2K
R6    7   8   2K
```

```
R7   8   9   2K
R8   9  10   2K
*    All diodes have the same model name of D1N914.
D1   1   5   D1N914
D2   1   4   D1N914
D3   1   3   D1N914
D4   1   7   D1N914
D5   1   8   D1N914
D6   1   9   D1N914
*    Diode model D1N914 defines diode parameters.
.MODEL D1N914 D (IS=100E-15 RS=16 BV=100 IBV=100E-15)
*    DC sweep from -15 V to 30 V with 0.5 V increment
.DC  VIN  -15  30  0.5
*    Plot the results of DC sweep: V(2) versus VIN.
.PLOT  DC  V(2)
*    Graphic post-processor
.PROBE
.END
```

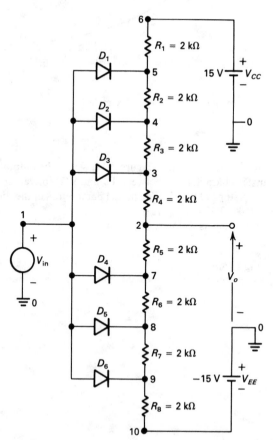

Figure 7.14 Diode waveform-shaping circuit

The results of the simulations are shown in Figure 7.15.

Example 7.6

A full-wave rectifier is shown in Figure 7.16, where the output is taken between terminals 4 and 3. Plot the transient response of the output voltage V(4, 3) for the time duration of 0 to 20 ms in steps of 0.1 ms. The peak voltage of the transformer primary is 120 V, 60 Hz. The turns ratio of primary to secondary windings is 10 : 1. The model parameters are the default values. Calculate and print the coefficients of Fourier series.

Solution   The transformer secondaries may be considered as a voltage-controlled voltage source, as shown in Figure 7.17. The primary is represented by a voltage source with a very high resistance, e.g., 10 MΩ.

The list of the circuit file is as follows.

```
Example 7.6     Rectifier with Single-phase Center-tapped Transformer
*    Primary is modeled as a voltage source of 120 V peak at 60 Hz
*    with zero offset voltage.
VP   1   0   SIN (0 120 60HZ)
*    Primary winding is assumed to have a very high resistance: R1 = 10 MΩ.
R1   1   0   10GOHM
*    Secondary winding is assumed as a voltage-controlled voltage source with
```

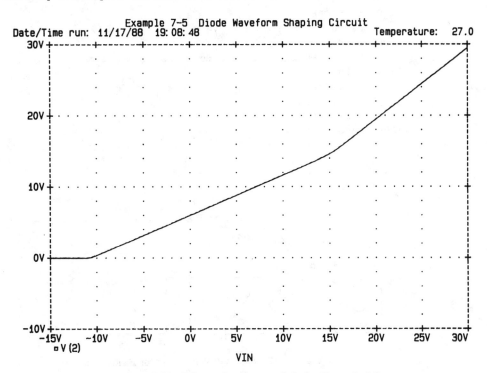

Figure 7.15   DC transfer characteristic for Example 7.5.

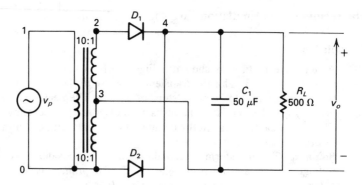

Figure 7.16 Rectifier with single-phase center-tapped transformer

```
*   a voltage gain of 0.1.
E1   2   3   1   0   0.1
E2   3   0   1   0   0.1
C1   4   3   50UF
RL   4   3   500
*   Diode D1 with model name DIODE
D1   2   4   DIODE
D2   0   4   DIODE
*   Diode model with default values
.MODEL DIODE D
*   Transient analysis from 0 to 20 ms with 0.1-ms increment
.TRAN   0.1MS 20MS
*   Plot the results of transient analysis for voltage across nodes
*   4 and 3.
.PLOT   TRAN   V(4,3)
.FOUR   60HZ   V(4,3)
*   Graphic post-processor
.PROBE
.END
```

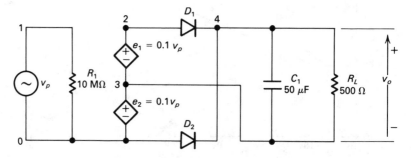

Figure 7.17 The equivalent circuit for Figure 7.16

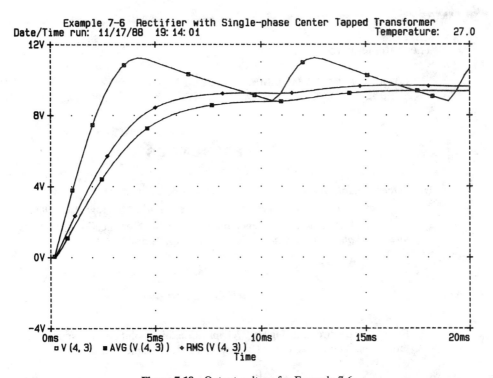

Figure 7.18 Output voltage for Example 7.6.

The transient response for Example 7.6 is shown in Figure 7.18. The coefficients of the Fourier analysis are given next.

```
****    FOURIER ANALYSIS                      TEMPERATURE =   27.000 DEG C
FOURIER COMPONENTS OF TRANSIENT RESPONSE V(4,3)
DC COMPONENT =   1.008665E+01
HARMONIC   FREQUENCY     FOURIER     NORMALIZED     PHASE        NORMALIZED
  NO         (HZ)       COMPONENT    COMPONENT      (DEG)       PHASE (DEG)
   1       6.000E+01    2.112E-03    1.000E+00    -2.607E+00     0.000E+00
   2       1.200E+02    9.843E-01    4.661E+02     1.182E+01     1.443E+01
   3       1.800E+02    7.890E-04    3.736E-01     2.452E+01     2.713E+01
   4       2.400E+02    3.833E-01    1.815E+02     2.131E+01     2.392E+01
   5       3.000E+02    1.167E-03    5.528E-01     8.314E+01     8.575E+01
   6       3.600E+02    1.669E-01    7.902E+01     3.761E+01     4.022E+01
   7       4.200E+02    1.756E-03    8.314E-01     1.076E+02     1.102E+02
   8       4.800E+02    7.140E-02    3.381E+01     6.687E+01     6.948E+01
   9       5.400E+02    1.895E-03    8.971E-01     1.279E+02     1.305E+02
   TOTAL HARMONIC DISTORTION =   5.075342E+04 PERCENT
       JOB CONCLUDED
       TOTAL JOB TIME          22.62
```

SUMMARY

The statements for diodes are

```
D⟨name⟩  NA   NK   DNAME   [(area) value]
.MODEL DNAME   D (P1=V1   P2=V2   P3=V3 ...PN=VN)
```

REFERENCES

1. *PSpice Manual.* Irvine, Calif.: MicroSim Corporation, 1988.
2. P. Antognetti, *Power Integrated Circuits.* New York: McGraw-Hill, 1986.
3. A. Laha and D. Smart, "A Zener diode model with application to SPICE2," *IEEE Journal of Solid-State Circuits,* Vol. SC-16, No. 1, pp. 21–22.

PROBLEMS

7.1. For the diode circuit in Figure P7.1, print the bias point and the small-signal parameters of the diode. Use default values of Model parameters.

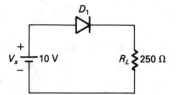

Figure P7.1

7.2. If the input voltage to the circuit in Figure 7.7 is v_{in} = 15 sin (2000π t), plot the transient response of the output voltage for a time duration of 0 to 2 ms with a time increment of 10 μs. Print the details of transient analysis bias point. The Zener voltages of the diodes are the same, V_Z = 5.2 V, and the current at the zener breakdown is I_Z = 0.5 μA. The model parameters are IS=0.5UA RS=6 CJO=2PF TT=12NS BV=5.20 IBV=0.5UA. The operating temperature is 50°C.

7.3. If the input voltage of the circuit in Figure 7.14 is V_{in} = 10 V DC, print the details of the DC operating point. Print the voltage gain (V_o/V_{in}), the input resistance, and the output resistance.

7.4. A full-wave bridge rectifier is shown in Figure P7.4. Plot the transient response of the output voltage for the time duration of 0 to 20 ms in steps of 0.1 ms. The model parameters are the default values. Print the details of transient analysis bias point and the coefficients of the Fourier series.

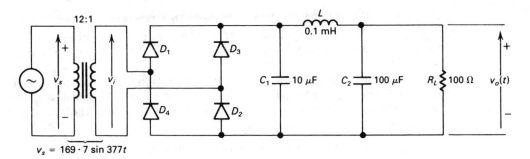

Figure P7.4

7.5. For the diode circuit in Figure P7.5, plot the DC transfer characteristic between v_{in} and v_o for values of v_{in} in the range of -18 V to 18 V in steps of 0.5 V. The model parameters of the diodes are the default values.

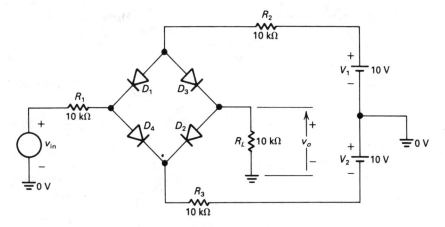

Figure P7.5

7.6. For the diode circuit in Figure P7.6, plot the input current against the input voltage for values of V_{in} in the range of -10 V to 10 V in steps of 0.25 V. The model parameters are IS=0.5UA RS=6 BV=5.20 IBV=0.5UA. The operating temperature is 50°C.

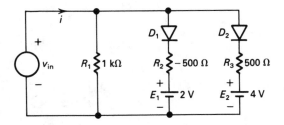

Figure P7.6

7.7. Repeat Example 7.3 if the direction of the diode D_1 is reversed.

7.8. Repeat Example 7.6 if the diodes are represented by voltage-controlled switches. The model parameters of the switches are RON=0.25 ROFF=1E+6 VON=0.25 VOFF=0.

7.9. A demodulator circuit is shown in Figure P7.9. Plot the transient output voltage for the time duration of 0 to 100 μs with an increment of 0.5 μs. The model parameters are IS=0.5UA RS=6 CJO=2PF TT=12NS BV=5.20 IBV=0.5UA. The input voltage is given by

$$v_i = 10[1 + 0.5 \sin(2\pi \times 10 \times 10^3 t)]\sin(2\pi \times 20 \times 10^6 t)$$

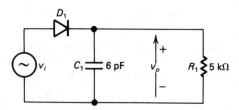

Figure P7.9

8

Bipolar Junction Transistors

8.1 INTRODUCTION

A bipolar junction transistor (BJT) may be specified by a device statement in conjunction with a model statement. Similar to diodes, the BJT model incorporates an extensive range of characteristics, e.g., DC and small-signal behavior, temperature dependency, and noise generation. The model parameters take into account temperature effects, various capacitances, and physical properties of semiconductors.

8.2 BJT MODEL

PSpice generates a complex model for BJTs. The model equations that are used by *PSpice* are described in [1] and [2]. If a complex model is not necessary, the model parameters can be ignored by the users, and *PSpice* assigns default values to the parameters.

The *PSpice* model, which is based on the integral charge-control model of Gummel and Poon [1, 6] is shown in Figure 8.1. The small-signal and static models that are generated by *PSpice* are shown in Figures 8.2 and 8.3, respectively.

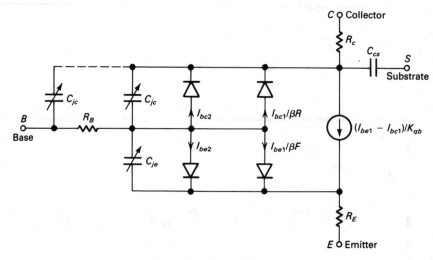

Figure 8.1   PSpice BJT model

The model statement for NPN transistors has the general form

```
.MODEL QNAME NPN (P1=V1 P2=V2 P3=V3 ...PN=VN)
```

and the general form for PNP transistors is

```
.MODEL QNAME PNP (P1=V1 P2=V2 P3=V3 ...PN=VN)
```

where QNAME is the name of the BJT model. NPN and PNP are the type

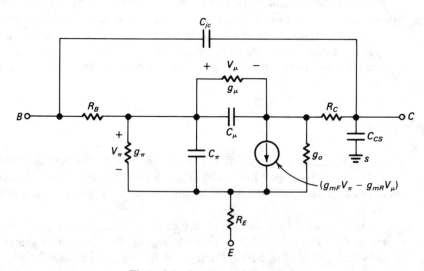

Figure 8.2   Small-signal BJT model

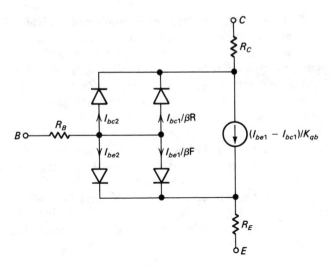

Figure 8.3 Static BJT model

symbols for NPN and PNP transistors, respectively. QNAME, which is the model name, can begin with any character, and its word size is normally limited to 8. P1, P2, . . . and V1, V2, . . . are the parameters and their values, respectively. Table 8.1 shows the model parameters of BJTs. If certain parameters are not specified, *PSpice* assumes the simple model of Ebers-Moll [3], which is shown in Figure 8.4(a).

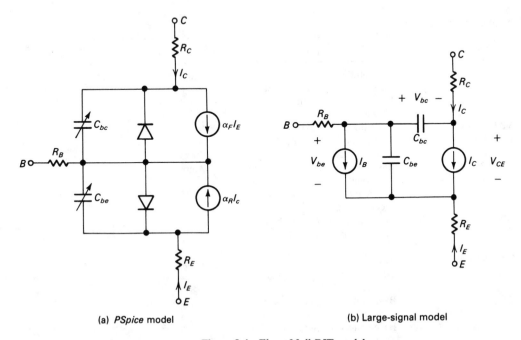

(a) *PSpice* model

(b) Large-signal model

Figure 8.4 Ebers-Moll BJT model

As with diodes, an *area factor* is used to determine the number of equivalent parallel BJTs of a specified model. The model parameters that are affected by the area factor are marked by an asterisk (*) in Table 8.1. A bipolar transistor is modeled as an intrinsic transistor with ohmic resistances in series with the collector (RC/area), the base (RB/area), and with the emitter (RE/area). [(area) value] is the relative device area and defaults to 1. For those parameters that have alternate names, such as VAF and VA (the alternate name is indicated with parentheses), either name may be used.

The parameters ISE (C2) and ISC (C4) may be set to be greater than 1. In this case, they are interpreted as multipliers of IS instead of absolute currents—that is, if ISE > 1, then it is replaced by ISE*IS, and likewise for ISC.

The DC model is defined (1) by parameters BF, C2, IK, and NE, which determine the forward current gain, (2) by BR, C4, IKR, and VC, which determine the reverse current gain characteristics, (3) by VA and VB, which determine the output conductance for forward and reverse regions, and (4) by the reverse saturation current IS.

Base-charge storage is modeled (1) by forward and reverse transit times TF and TF and nonlinear depletion layer capacitances, which are determined by CJE, PE, and ME for b-e junction, and (2) by CJC, PC, and MC for b-c junction. CCS is a constant collector-substrate capacitance.

The temperature dependence of the saturation current is determined by the energy gap EG and the saturation current temperature exponent PT.

TABLE 8.1 MODEL PARAMETERS OF BJTS

Name	Area	Model parameters	Units	Default	Typical
IS	*	pn saturation current	Amps	1E-16	1E-16
BF		Ideal maximum forward beta		100	100
NF		Forward current emission coefficient		1	1
VAF(VA)		Forward Early voltage	Volts	∞	100
IKF(IK)		Corner for forward beta high-current roll-off	Amps	∞	10M
ISE(C2)		Base-emitter leakage saturation current	Amps	0	1000
NE		Base-emitter leakage emission coefficient		1.5	2
BR		Ideal maximum reverse beta		1	0.1
NR		Reverse current emission coefficient		1	
VAR(VB)		Reverse early voltage	Volts	∞	100
IKR	*	Corner for reverse beta high-current roll-off	Amps	∞	100M
ISC(C4)		Base-collector leakage saturation current	Amps	0	1
NC		Base-collector leakage emission coefficient		2	2

TABLE 8.1   CONTINUED

Name	Area	Model parameters	Units	Default	Typical
RB	*	Zero-bias (maximum) base resistance	Ohms	0	100
RBM		Minimum base resistance	Ohms	RB	100
IRB		Current at which RB falls halfway to RBM	Amps	∞	
RE	*	Emitter ohmic resistance	Ohms	0	1
RC	*	Collector ohmic resistance	Ohms	0	10
CJE	*	Base-emitter zero-bias pn capacitance	Farads	0	2P
VJE(PE)		Base-emitter built-in potential	Volts	0.75	0.7
MJE(ME)		Base-emitter pn grading factor	0.33	0.33	
CJC	*	Base-collector zero-bias pn capacitance	Farads	0	1P
VJC(PC)		Base-collector built-in potential	Volts	0.75	0.5
MJC(MC)		Base-collector pn grading factor	0.33	0.33	
XCJC		Fraction of C_{bc} connected internal to R_B		1	
CJS(CCS)		Collector-substrate zero-bias pn capacitance	Farads	0	2PF
VJS(PS)		Collector-substrate built-in potential	Volts	0.75	
MJS(MS)		Collector-substrate pn grading factor		0	
FC		Forward-bias depletion capacitor coefficient		0.5	
TF		Ideal forward transit time	Seconds	0	0.1NS
XTF		Transit time bias dependence coefficient		0	
VTF		Transit time dependency on V_{bc}	Volts	∞	
ITF		Transit time dependency on I_c	Amps	0	
PTF		Excess phase at $1/(2\pi*TF)$Hz	Degrees	0	30°
TR		Ideal reverse transit time	Seconds	0	10NS
EG		Bandgap voltage (barrier height)	Electron-volts	1.11	1.11
XTB		Forward and reverse beta temperature coefficient		0	
XTI(PT)		IS temperature effect exponent		3	
KF		Flicker noise coefficient		0	6.6E-16
AF		Flicker noise exponent		1	1

8.3 BJT STATEMENTS

The symbol for a bipolar junction transistor (BJT) is Q. The name of a bipolar transistor must start with Q and it takes the general form

```
Q⟨name⟩   NC  NB  NE  NS  QNAME  [⟨area⟩ value]
```

where NC, NB, NE, and NS are the collector, base, emitter, and substrate nodes, respectively. QNAME could be any name of up to 8 characters. The substrate node is optional; if not specified it defaults to ground. Positive current is the current that flows into a terminal. That is, the current flows from the collector node through the device to the emitter node for an NPN BJT.

Some Statements for BJTs

```
QIN   5   7   8     2N2222
QS    2   4   5     2N2907    1.5
QX    1   4   9     NMOD
.MODEL 2N2222 NPN(IS=3.108E-15 XTI=3 EG=1.11 VAF=131.5 BF=217.5
+   NE=1.541 ISE=190.7E-15 IKF=1.296 XTB=1.5 BR=6.18 NC=2 ISC=0 IKR=0
+   RC=1 CJC=14.57E-12 VJC=.75 MJC=.3333 FC=.5 CJE=26.08E-12 VJE=.75
+   MJE=.3333 TR=51.35E-9 TF=451E-12 ITF=.1 VTF=10 XTF=2)
.MODEL 2N2907 PNP(IS=9.913E-15 XTI=3 EG=1.11 VAF=90.7 BF=197.8
+   NE=2.264 ISE=6.191E-12 IKF=.7322 XTB=1.5 BR=3.369 NC=2 ISC=0 IKR=0
+   RC=1 CJC=14.57E-12 VJC=.75 MJC=.3333 FC=.5 CJE=20.16E-12 VJE=.75
+   MJE=.3333 TR=29.17E-9 TF=405.7E-12 ITF=.4 VTF=10 XTF=2)
.MODEL NMOD NPN
```

Note. A + (plus) sign at the first column indicates the continuation of the statement preceding it.

Example 8.1

A bipolar transistor circuit is shown in Figure 8.5(a), where the output is taken from node 4. Calculate and print the sensitivity of collector current with respect to all parameters. Print the details of the bias point. The equivalent circuit for transistor Q_1 is shown in Figure 8.5(b).

Solution   The list of the circuit file is as follows.

Example 8.1   Biasing Sensitivity of Bipolar Transistor Amplifier
```
.OPTIONS NOPAGE NOECHO
*   Sensitivity of collector current (which is the current through voltage
*   source VRC)
.SENS  I(VRC)
*  Supply voltage is 15 V DC.
VCC  7  0   DC 15V
*   A dummy voltage source of 0 V to measure the collector current
VRC  6  4   DC  0V
R1   7  3   47K
R2   3  0   2K
RC   7  6   10K
RE   5  0   2K
*   Subcircuit call for transistor model QMOD and the substrate is
*   connected to ground by default.
XQ1 4  3  5   QMOD
```

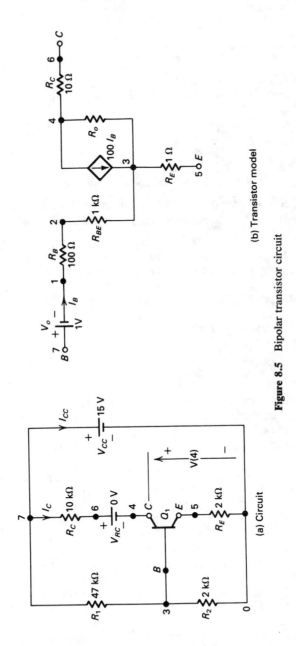

(b) Transistor model

(a) Circuit

Figure 8.5 Bipolar transistor circuit

```
*    Subcircuit definition for QMOD
.SUBCKT QMOD  6  7  5
RB   1  2  100
RE   3  5  1
RC   4  6  10
RBE  2  3  1K
RO   4  3  100K
*    A dummy voltage source of 0 V to measure the controlling current
VI   7  1  DC 0V
F1   4  3  VI  20
*    End of subcircuit definition
.ENDS  QMOD
.END
```

The .SENS command does not require a .PRINT command for printing the output. The output of sensitivity analysis and the bias point are given next.

```
****      SMALL-SIGNAL BIAS SOLUTION        TEMPERATURE =   27.000 DEG C
NODE    VOLTAGE        NODE    VOLTAGE      NODE    VOLTAGE      NODE    VOLTAGE
(    3)     .5960    (    4)   12.1520    (    5)     .5864    (    6)   12.1520
(    7)   15.0000    (XQ1.1)     .5960    (XQ1.2)     .5952    (XQ1.3)     .5867
(XQ1.4)   12.1500
      VOLTAGE SOURCE CURRENTS
      NAME         CURRENT
      VCC          -5.912E-04
      VRC          2.848E-04
      XQ1.VI       8.456E-06
      TOTAL POWER DISSIPATION   8.87E-03  WATTS
****      DC SENSITIVITY ANALYSIS            TEMPERATURE =   27.000 DEG C
DC SENSITIVITIES OF OUTPUT I(VRC)
          ELEMENT        ELEMENT        ELEMENT        NORMALIZED
          NAME           VALUE          SENSITIVITY    SENSITIVITY
                                        (AMPS/UNIT)    (AMPS/PERCENT)
          R1             4.700E+04      -5.481E-09     -2.576E-06
          R2             2.000E+03       1.252E-07      2.505E-06
          RC             1.000E+04      -3.134E-10     -3.134E-08
          RE             2.000E+03      -1.288E-07     -2.576E-06
          XQ1.RB         1.000E+02      -3.705E-09     -3.705E-09
          XQ1.RE         1.000E+00      -1.288E-07     -1.288E-09
          XQ1.RC         1.000E+01      -3.134E-10     -3.134E-11
          XQ1.RBE        1.000E+03      -3.705E-09     -3.705E-08
          XQ1.RO         1.000E+05      -1.273E-10     -1.273E-07
          VCC            1.500E+01       1.898E-05      2.848E-06
          VRC            0.000E+00      -1.101E-06      0.000E+00
          XQ1.VI         0.000E+00      -4.381E-04      0.000E+00
      JOB CONCLUDED
      TOTAL JOB TIME              2.52
```

Example 8.2

A bipolar Darlington pair amplifier is shown in Figure 8.6. Calculate and print the voltage gain, the input resistance, and the output resistance. The input voltage is 5 V. The model parameters of the bipolar transistors are BF=100 BR=1 RB=5 RC=1 RE=0 VJE=0.8 VA=100.

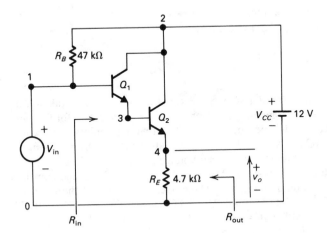

Figure 8.6 Darlington pair amplifier

Solution The list of the circuit file is as follows.

```
Example 8.2   Darlington Pair
.OPTIONS   NOPAGE   NOECHO
VCC   2   0   DC   12V
VIN   1   0   DC   5V
*     BJTs with model QM
Q1    2   1   3   QM
Q2    2   3   4   QM
RB    2   1   47k
RE    4   0   4.7K
*     Model QM for NPN BJTs
.MODEL QM   NPN (BF=100 BR=1 RB=5 RC=1 RE=0 VJE=0.8 VA=100)
*     Transfer function analysis to calculate DC gain, input
*     resistance, and output resistance
.TF   V(4)   VIN
.END
```

The results of the transfer function analysis are given next.

```
****      SMALL-SIGNAL BIAS SOLUTION       TEMPERATURE =   27.000 DEG C
 NODE   VOLTAGE       NODE   VOLTAGE      NODE    VOLTAGE      NODE   VOLTAGE
(    1)   5.0000  (     2)   12.0000  (    3)    4.3560  (    4)    3.5909
```

```
VOLTAGE SOURCE CURRENTS
NAME          CURRENT
VCC          -9.129E-04
VIN           1.489E-04
TOTAL POWER DISSIPATION   1.02E-02   WATTS
****      SMALL-SIGNAL CHARACTERISTICS
   V(4)/VIN =   9.851E-01
   INPUT RESISTANCE AT VIN =   4.696E+04
   OUTPUT RESISTANCE AT V(4) =   6.679E+01
      JOB CONCLUDED
      TOTAL JOB TIME              2.97
```

Example 8.3

A bipolar transistor amplifier circuit is shown in Figure 8.7. The output is taken from node 6. Calculate and plot the magnitude and phase of the voltage gain for frequencies from 1 Hz to 10 kHz with a decade increment and with 10 points per decade. The input voltage for AC analysis is 10 mV. Calculate and plot the transient response of voltages at nodes 4 and 6 for an input voltage of $v_{in} = 0.01 \sin(2\pi \times 1000t)$ and for a duration of 0 to 2 ms insteps of 50 μs. The details of AC and transient analysis operating points should be printed. The model parameters of the PNP BJT are IS=2E-16 BF=50 BR=1 RB=5 RC=1 RE=0 TF=0.2NS TR=5NS CJE=0.4PF VJE=0.8 ME=0.4 CJC=0.5PF VJC=0.8 CCS=1PF VA=100.

Solution   The list of the circuit file is as follows.

Example 8.3   Bipolar Transistor Amplifier

```
.OPTIONS   NOPAGE NOECHO
*    Transient analysis for 0 to 2 ms with 50-μs increment
*    Print details of transient analysis operating point.
.TRAN/OP  50US  2MS
*    AC analysis from 1 Hz to 10 KHz with a decade increment and
*    10 points per decade
```

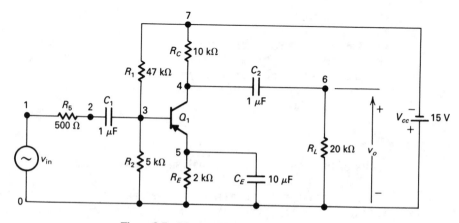

Figure 8.7   Bipolar transistor amplifier circuit

```
.AC  DEC  10  1HZ  10KHZ
*  Print the details of AC analysis operating point.
.OP
*  Input voltage is 10 mV peak for ac analysis and for transient response:
*  It is 10 mV peak at 1 kHz with zero-offset value.
VIN  1  0  AC  10MV  SIN(0  10MV  1KHZ)
VCC  0  7  DC  15V
RS   1  2  500
R1   7  3  47K
R2   3  0  5K
RC   7  4  10K
RE   5  0  2K
RL   6  0  20K
C1   2  3  1UF
C2   4  6  1UF
CE   5  0  100UF
*  Transistor Q1 with model QM
Q1   4  3  5  0  QM
*  Model QM for PNP transistors
.MODEL  QM  PNP (IS=2E-16 BF=50 BR=1 RB=5 RC=1 RE=0 TF=0.2NS TR=5NS
+         CJE=0.4PF VJE=0.8 ME=0.4 CJC=0.5PF VJC=0.8 CCS=1PF VA=100)
*  Plot the results of transient analysis for voltages at nodes 4, 6, and 1
.PLOT  TRAN  V(4) V(6) V(1)
*  Plot the results of AC analysis for the magnitude and phase angle
*  of output voltage at node 6
.PLOT  AC  VM(6)  VP(6)
.PROBE
.END
```

The determination of the operating point is the first step in analyzing a circuit with nonlinear devices (e.g., bipolar transistors). The equivalent circuit for determining the AC analysis (or DC analysis) bias point of the amplifier in Figure 8.7 is shown in Figure 8.8, where the capacitors are open-circuited. The details of the bias point are given next.

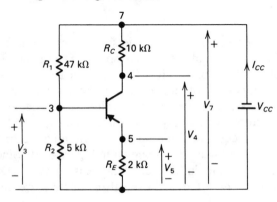

Figure 8.8 Equivalent circuit for DC bias calculation

```
****      SMALL-SIGNAL BIAS SOLUTION      TEMPERATURE =   27.000 DEG C
  NODE    VOLTAGE      NODE    VOLTAGE      NODE   VOLTAGE      NODE    VOLTAGE
(   1)     0.0000   (    2)     0.0000   (    3)   -1.4280   (    4)  -11.5240
(   5)    -.7016    (    6)     0.0000   (    7)  -15.0000
     VOLTAGE SOURCE CURRENTS
     NAME          CURRENT
     VIN           0.000E+00
     VCC          -6.364E-04
     TOTAL POWER DISSIPATION   9.55E-03   WATTS
```

Once the DC bias point is determined, *PSpice* generates a small-signal model of
the BJT. This model is similar to that in Figure 8.9. *PSpice* replaces the transistor
by this circuit model. It should be noted that this model is valid only at the
operating point. The details of the operating point are given next.

```
   ****      OPERATING POINT INFORMATION       TEMPERATURE =   27.000 DEG C
   **** BIPOLAR JUNCTION TRANSISTORS
   NAME          Q1
   MODEL         QM
   IB           -3.16E-06
   IC           -3.48E-04
   VBE          -7.26E-01
   VBC           1.01E+01
   VCE          -1.08E+01
   BETADC        1.10E+02
   GM            1.34E-02
   RPI           8.19E+03
   RX            5.00E+00
   RO            3.17E+05
   CBE           3.39E-12
   CBC           2.11E-13
   CBX           0.00E+00
   CJS           1.00E-12
   BETAAC        1.10E+02
   FT            5.94E+08
```

Prior to the transient analysis, *PSpice* determines the small-signal parameters of
the nonlinear devices and the potentials of the various nodes. The method for the
calculation of transient analysis bias point differs from that of DC analysis bias
point because, in transient analysis, all the nodes have to be assigned the initial
values and the nonlinear sources may have transient values at the beginning of
transient analysis. The capacitors, which may have initial values, therefore remain as parts of the circuit. The equivalent circuit for determining the transient
analysis bias point for the circuit in Figure 8.7 is shown in Figure 8.10. Since the
capacitors in Figure 8.7 do not have any initial values, the bias points for DC and
transient analysis are the same. There the small-signal parameters are also the

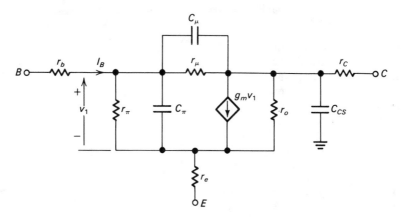

Figure 8.9   Small-signal equivalent circuit of bipolar transistors

same. The details of the transient analysis boas point and the small-signal parameters are given next to compare with that of DC analysis.

```
****      INITIAL TRANSIENT SOLUTION         TEMPERATURE =   27.000 DEG C
 NODE    VOLTAGE        NODE    VOLTAGE       NODE    VOLTAGE      NODE    VOLTAGE
(   1)    0.0000   (    2)     0.0000   (     3)    -1.4280   (    4)   -11.5240
(   5)   -.7016    (    6)     0.0000   (     7)   -15.0000
    VOLTAGE SOURCE CURRENTS
    NAME          CURRENT
    VIN           0.000E+00
    VCC          -6.364E-04
    TOTAL POWER DISSIPATION    9.55E-03  WATTS
****      OPERATING POINT INFORMATION        TEMPERATURE =   27.000 DEG C
**** BIPOLAR JUNCTION TRANSISTORS
```

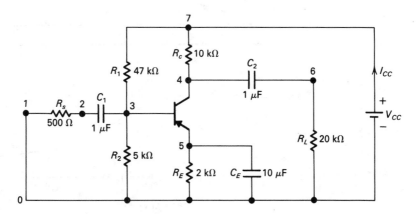

Figure 8.10   Equivalent circuit for transient analysis bias point

```
NAME          Q1
MODEL         QM
IB            -3.16E-06
IC            -3.48E-04
VBE           -7.26E-01
VBC           1.01E+01
VCE           -1.08E+01
BETADC        1.10E+02
GM            1.34E-02
RPI           8.19E+03
RX            5.00E+00
RO            3.17E+05
CBE           3.39E-12
CBC           2.11E-13
CBX           0.00E+00
CJS           1.00E-12
BETAAC        1.10E+02
FT            5.94E+08
```

The frequency and transient responses are shown in Figures 8.11 and 8.12, respectively.

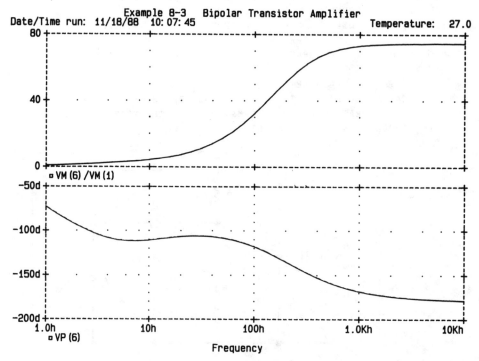

Figure 8.11   Frequency response for Example 8.3

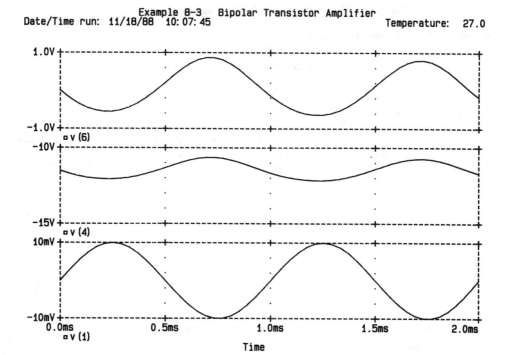

Figure 8.12   Transient response for Example 8.3

Example 8.4

If the transistor in Figure 8.7 is replaced by an equivalent circuit in Figure 8.13, repeat Example 8.3. There is no need to print the details of operating point.

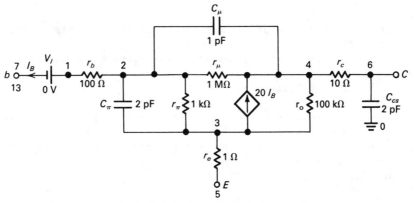

Figure 8.13   Subcircuit for PNP bipolar transistor

Solution   The list of the circuit file is as follows.

Example 8.4   Bipolar Transistor Amplifier

```
.OPTIONS  NOPAGE  NOECHO
*   Transient analysis for 0 to 2 ms with 50-µs increment
.TRAN  50US  2MS
*   AC analysis from 1 Hz to 10 KHz with a decade increment and
*   10 points per decade
.AC  DEC  10  1HZ  10KHZ
*   Input voltage is 10 mV peak for ac analysis and for transient response:
*   It is 10 mV peak at 1 kHz with zero-offset value.
VIN  1  0  AC  1  SIN(0 0.01 1KHZ)
VCC  0  7  DC  15V
RS  1  2  500
R1  7  3  47K
R2  3  0  2K
RC  7  4  10K
RE  5  0  2K
RL  6  0  20K
C1  2  3  1UF
C2  4  6  1UF
CE  5  0  100UF
*   Calling subcircuit for transistor model TRANS
XQ1  4  3  5  TRANS
*   Subcircuit definition for TRANS
.SUBCKT  TRANS  6  7  5
RB  1  2  100
RE  3  5  1
RC  4  6  10
RPI  2  3  1K
CPI  2  3  2PF
RU  2  4  1MEG
CU  2  4  1PF
RO  4  3  100K
CCS  6  0  2PF
*   A dummy voltage source of 0 V through which the controlling current flows
VI  1  7  DC  0V
*  The collector current is controlled by the current through source VI.
F1  3  4  VI  20
*   End of subcircuit definition
.ENDS  TRANS
*   Plot the results of transient analysis for voltages at nodes 4, 6 and 1.
.PLOT  TRANS  V(4)  V(6)  V(1)
*   Plot the results of ac analysis for the magnitude and phase angle
*   of voltage at node 6.
.PLOT  AC  VM(6)  VP(6)
.PROBE
.END
```

The frequency and transient responses are shown in Figures 8.14 and 8.15, respectively.

Example 8.5

A two-stage bipolar transistor amplifier is shown in Figure 8.16. The output is taken from node 9. Plot (a) the magnitude and phase angle of the voltage gain and (b) the magnitude of input impedance for frequencies from 10 Hz to 10 MHZ with a decade increment and 10 points per decade. The peak input voltage is 1 mV. The model parameters of the BJTs are IS=2E-16 BF=50 BR=1 RB=5 RC=1 RE=0 CJE=0.4PF VJE=0.8 ME=0.4 CJC=0.5PF VJC=0.8 CCS=1PF VA=100.
Solution   The list of the circuit file is as follows.

```
Example 8.5    Two-Stage BJT Amplifier
VCC 10   0   DC   15V
*   Input voltage is 1 mV peak for frequency response.
VIN  1   0   AC   1MV
*   A dummy voltage source of 0 V to measure the input current
```

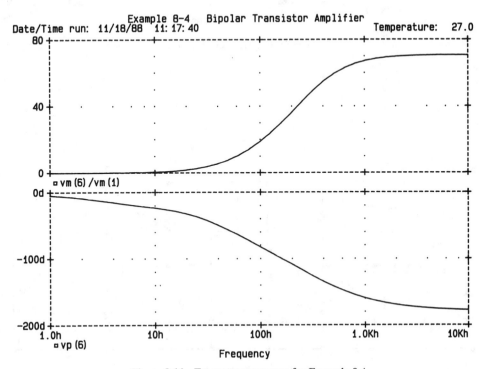

Figure 8.14   Frequency response for Example 8.4

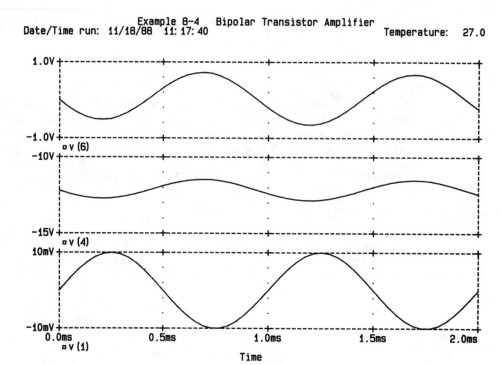

Figure 8.15 Transient response for Example 8.4

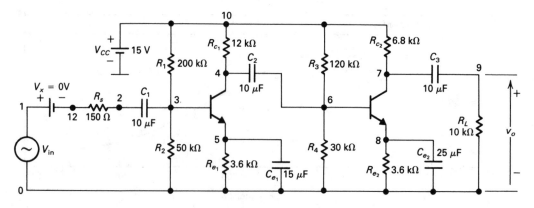

Figure 8.16 Two-Stage BJT Amplifier

```
VX    1    12   DC   0V
RS    12   2    150
C1    2    3    10UF
R1    10   3    200K
R2    3    0    50K
*  Transistors Q1 and Q2 have model QM.
Q1    4    3    5    0    QM
Q2    7    6    8    0    QM
RC1   10   4    12K
RE1   5    0    3.6K
CE1   5    0    15UF
C2    4    6    10UF
R3    10   6    120K
R4    6    0    30K
RC2   10   7    6.8K
RE2   8    0    3.6K
CE2   8    0    25UF
C3    7    9    10UF
RL    9    0    10K
*  Model statement for NPN transistors whose model name is QM
.MODEL  QM  NPN (IS=2E-16 BF=50 BR=1 RB=5 RC=1 RE=0 CJE=0.4PF
+               VJE=0.8 ME=0.4 CJC=0.5PF VJC=0.8 CCS=1PF VA=100)
*  AC analysis from 10 Hz to 10 MHz with a decade increment and 10
*  points per decade
.AC  DEC  10  10HZ  10MEGHZ
.PLOT  AC  VM(9)  VP(9)
.PROBE
.END
```

The results of the frequency response are shown in Figure 8.17.

Example 8.6

A two-stage amplifier with shunt-series feedback is shown in Figure 8.18. Plot (a) the magnitude and phase angle of voltage gain and (b) the magnitude of the input impedance if the frequency is varied from 100 Hz to 100 MHz in steps of decade and 10 points per decade. The peak input voltage is 10 mV. The model parameters of the BJTs are IS=2E-16 BF=50 BR=1 RB=5 RC=1 RE=0 CJE=0.4PF VJE=0.8 ME=0.4 CJC=0.5PF VJC=0.8 CCS=1PF VA=100.

Solution

```
Example 8.6    Two-stage BJT amplifier with shunt-series feedback
VCC 10   0 DC 15V
*  Input voltage of 10 mV peak for frequency response
VIN  1   0  AC  10MV
*  A dummy voltage source of 0 V
VX   1   12  DC  0V
```

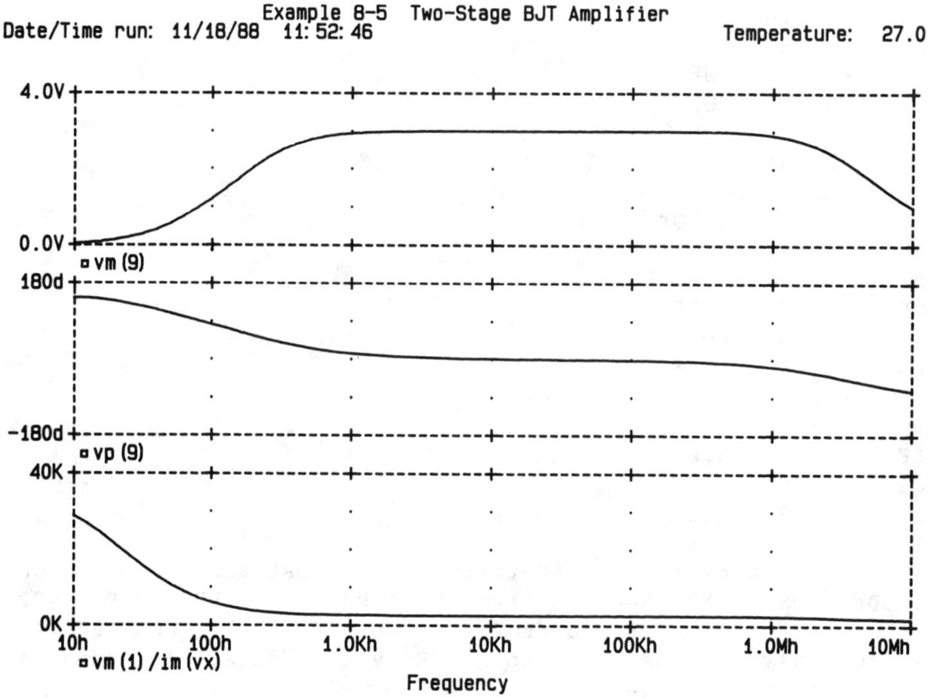

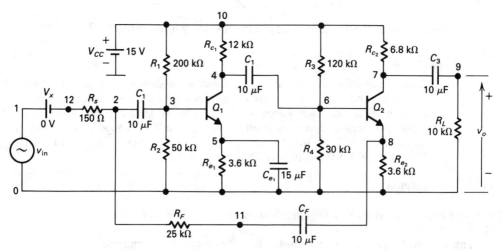

Figure 8.17   Frequency response for Example 8.5

Figure 8.18   Two-stage BJT amplifier with shunt-series feedback

```
RS    12   2    150
C1    2    3    10UF
R1    10   3  200K
R2    3    0   50K
*   Substrate of BJTs with model QM is connected to node 0.
Q1    4    3    5    0   QM
Q2    7    6    8    0   QM
RC1   10   4    1.2K
RE1   5    0    3.6K
CE1   5    0    15UF
C2    4    6    10UF
R3    10   6    120K
R4    6    0    30K
RC2   10   7    6.8K
RE2   8    0    3.6K
CF    11   8    10UF
RF    2    11   25K
C3    7    9    10UF
RL    9    0    10K
*   Model statement for NPN transistors with model name QM
.MODEL   QM NPN (IS=2E-16 BF=50 BR=1 RB=5 RC=1 RE=0 CJE=0.4PF
+               VJE=0.8 ME=0.4 CJC=0.5PF VJC=0.8 CCS=1PF VA=100)
*   AC analysis for 10 Hz to 100 MHz with a decade increment and 10
*   points per decade
.AC  DEC  10   10   10MEGHZ
.PLOT  AC  VM(9)  VP(9)
.END
```

The results of the frequency response are shown in Figure 8.19.

Example 8.7

An astable multivibrator is shown in Figure 8.20. The output is taken from nodes 1 and 2. Plot the transient responses of voltages at nodes 1 and 2 from 0 to 10 μs in steps of 0.01 μs. The initial voltages of nodes 1 and 3 are 0. The CPU time should be limited to 1.22E2 s. The model parameters of the BJTs are IS=2E-16 BF=50 BR=1 RB=5 RC=1 RE=0 TF=0.2NS TR=5NS.

Solution Due to the regenerative nature of the circuit, the solution may not converge and the simulation will continue for a very long time. The CPU time is limited so that the circuit does not run for a long time. The run time should be less than the CPU time itself if the circuit convergences. The list of the circuit file is as follows.

```
Example 8.7    Astable Multivibrator
VCC   6    0    DC   5V
RC1   6    1    1K
RC2   6    2    1K
```

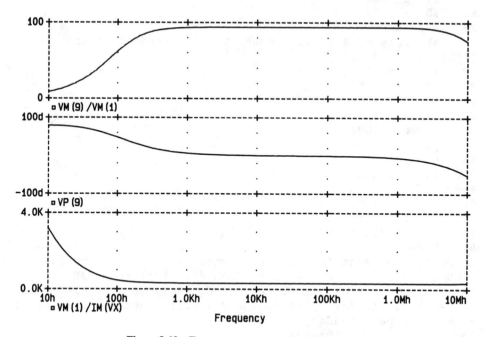

Example 8-6 Two-stage BJT amplifier with shunt-series feedback
Date/Time run: 11/18/88 14:28:34   Temperature: 27.0

Figure 8.19 Frequency response for Example 8.6

```
R1      6    3    30K
R2      6    4    30K
C1      1    4    150PF
C2      2    3    150PF
*   Q1 and Q2 with model QM and substrate connected to ground by
+ default
Q1      1    3    0    QM
Q2      2    4    0    QM
```

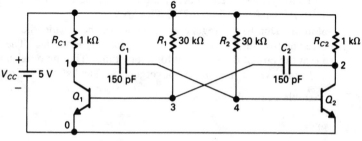

Figure 8.20 Astable multivibrator

```
*   Model statement for NPN transistors
.MODEL QM NPN (IS=2E-16 BF=50 BR=1 RB=5 RC=1 RE=0 TF=0.2NS TR=5NS)
*   CPU time is limited.
.OPTIONS  NOPAGE  NOECHO  CPTIME=1.2E2
*   Node voltages are set to defined values to break the tie-in
condition.
.NODESET V(1)=0  V(3)=0
*   Transient analysis from 0 to 10 μs with 0.1-μs increment
.TRAN/OP  0.1US  10US
*   Plot the results of transient analysis: voltages at nodes 2 and
+4.
.PLOT  TRAN  V(1)  V(2)
.PROBE
.END
```

The transient responses are shown in Figure 8.21.

Example 8.8

A TTL inverter circuit is shown in Figure 8.22(a). The output is taken from node 4. Plot the dc transfer characteristic V(4) versus V_{in} if the input voltage is varied from 0

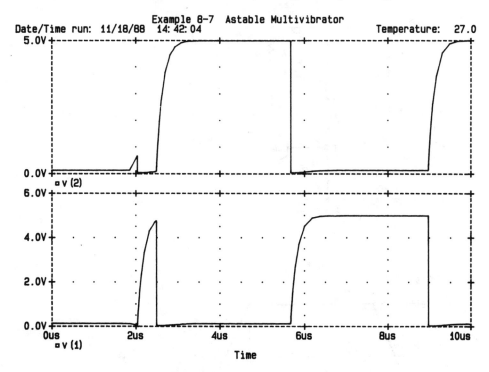

Figure 8.21 Transient responses for Example 8.7

to 2 V with a step of 0.01 V. If the input is a pulse voltage with a period of 60 μs, as shown in Figure 8.22(b), plot the transient response of voltage at node 4 from 0 to 80 ns in steps of 1 ns. The model parameters of the BJTs are BF=50 RB=70 RC=40 CCS=2PF TF=0.1NS TR=10NS VJC=0.85 VAF=50.
Solution   The list of the circuit file is as follows.

Example 8.8   TTL Inverter

```
*    Pulsed input voltage
VIN   1   0   PULSE (0   5   1NS   1NS   1NS   38NS   60NS)
VCC   6   0   DC   5V
*    BJTs with model QN and substrate connected to ground by default
Q1    3   2   1   QN
Q2    4   3   5   QN
Q3    4   5   0   QN
*    Model for NPN BJTs with model QN
.MODEL QN NPN (BF=50  RB=70  RC=40  CCS=2PF  TF=0.1NS  TR=10NS  VJC=0.85
+VAF=50)
```

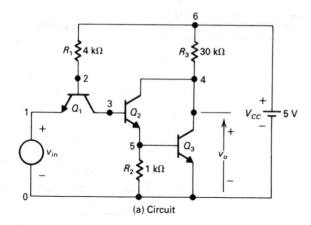

(a) Circuit

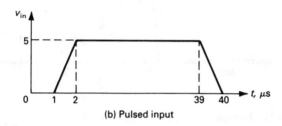

(b) Pulsed input

Figure 8.22   A TTL inverter

```
R1    6   2   4K
R2    5   0   1K
R3    6   4   1K
*   DC sweep for 0 to 2 with 0.01 V increment
.DC  VIN  0   2   0.01
*   Transient analysis for 0 to 80 ns with 1-ns increment
.TRAN  1NS   80NS
*   Plot the results of DC sweep: voltage at node 4 versus VIN.
.PLOT  DC  V(4)
*   Plot the results of transient analysis: Voltage at nodes 4 and 1.
.PLOT  TRAN  V(4)  V(1)
.PROBE
.END
```

The results of the DC sweep and transient analyses are shown in Figures 8.23 and 8.24, respectively.

Example 8.9

A TTL inverter circuit is shown in Figure 8.25(a). Plot the DC transfer characteristic between nodes 1 and 9 for values of V_{in} in the range of 0 to 2 V insteps of 0.01 V. If

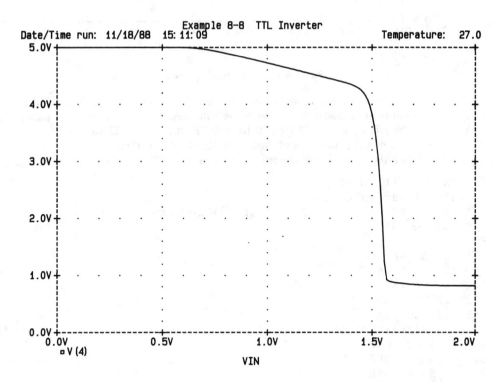

Figure 8.23   DC transfer characteristic for Example 8.8

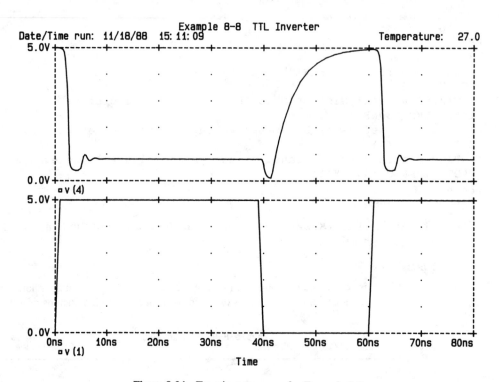

Figure 8.24 Transient response for Example 8.8

the input is a pulsed waveform of period 80 μs, as shown in Fig. 8.25(b), plot the transient response from 0 to 80 ns with steps of 1 ns. The model parameters of the BJTs are BF=50 RB=70 RC=40 TF=0.1NS TR=10NS VJC=0.85 VAF=50. The model parameters of diodes are RS=40 TT=0.1NS.

Solution   The list of the circuit file is as follows.

```
Example 8.9     TTL Inverter
*   Pulse input voltage
VIN   1   0   PULSE (0   3.5V   1NS   1NS   1NS   38NS   80NS)
VCC   13   0   5V
RS    1   2   50
RB1   13   3   4K
RC2   13   5   1.4K
RE2   6   0   1K
RC3   13   7   100
RB5   13   10   4K
*   BJTs with model QNP and substrate connected to ground by default.
Q1    4   3   2   QNP
Q2    5   4   6   QNP
Q3    7   5   8   QNP
Q4    9   6   0   QNP
Q5    11   10   9   QNP
```

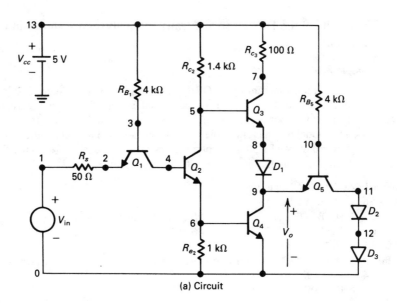

(a) Circuit

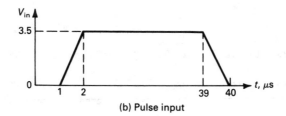

(b) Pulse input

Figure 8.25   TTL inverter

```
*   Diodes with model DIODE
D1   8   9    DIODE
D2   11  12   DIODE
D3   12  0    DIODE
*   Model of NPN transistors with model QNP
.MODEL QNP NPN (BF=50 RB=70 RC=40 TF=0.1NS TR=10NS VJC=0.85 VAF=50)
*   Diodes with model DIODE
.MODEL DIODE D (RS=40 TT=0.1NS)
*   DC sweep from 0 to 2 V with 0.01 V increment
.DC  VIN  0  2  0.01
*   Transient analysis from 0 to 80 ns with 1-ns increment
.TRAN  1NS  80NS
*   Plot the results of dc sweep: voltage at node 9 against VIN.
.PLOT  DC  V(9)
*   Plot the results of transient analysis: voltage at node 9.
.PLOT  TRAN  V(9)
.PROBE
.END
```

The results of the DC sweep and transient analyses are shown in Figures 8.26 and 8.27, respectively.

Example 8.10

The circuit diagram of an OR/NOR gate is shown in Figure 8.28(a). The inputs to nodes 1 and 4 are pulses of period 60 μs, as shown in Figure 8.28(b). Plot the transient responses of voltages at nodes 12, 13, and 1 from 0 to 80 ns in steps of 1 ns. The model parameters of the BJTs are BF=50 RB=70 RC=40 TF=0.1NS TR=10NS VJC=0.85 VAF=50. The parameters of the diodes are RS=40 TT=0.1NS.

Solution The list of the circuit file is as follows.

```
Example 8.10   OR/NOR Logic Gate
*   Pulsed input voltages
VA   1   0   PULSE (0   -5   1NS   1NS   1NS   38NS   60NS)
VB   4   0   PULSE (0   -5   1NS   1NS   1NS   38NS   60NS)
VEE  0   14  DC   5.2V
*   BJTs with model QN and substrate connected to ground by default
Q1   5   4   3   QN
Q2   7   8   3   QN
Q3   2   1   3   QN
Q4   0   9   8   QN
Q5   0   2   13  QN
Q6   0   7   12  QN
.MODEL QN NPN (BF=50 RB=70 RC=40 TF=0.1NS TR=10NS VJC=0.85 VAF=50)
```

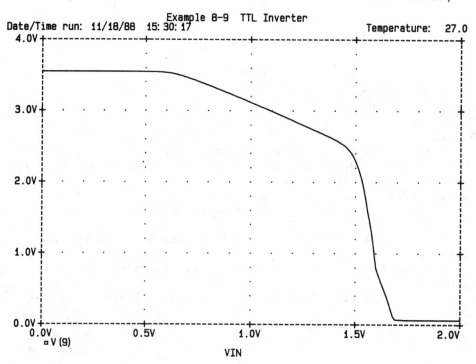

Figure 8.26 DC transfer characteristic for Example 8.9

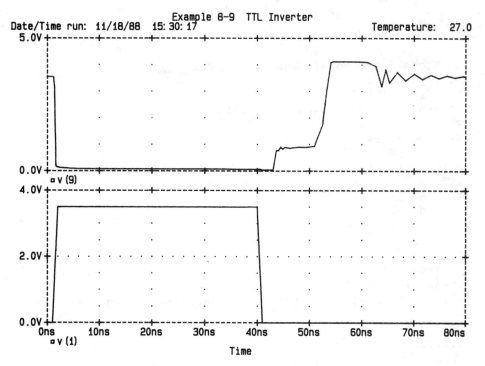

Figure 8.27 Transient response for Example 8.9

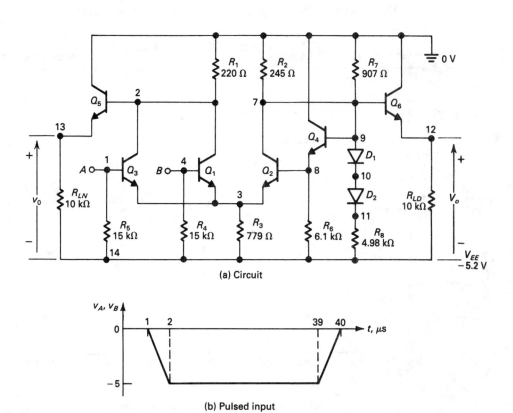

(a) Circuit

(b) Pulsed input

Figure 8.28 OR/NOR logic gate

```
*  Diodes with model DIODE
D1   9  10  DIODE
D2   10  11  DIODE
.MODEL DIODE  D (RS=40  TT=0.1NS)
R1   0  2  220
R2   0  7  245
R3   3  14  779
R4   4  14  15K
R5   1  14  15K
R6   8  14  6.1K
R7   0  9  907
R8   11  14  4.98K
RLO  12  14  10K
RLN  13  14  10K
*  Transient analysis from 0 to 80 ns with 1-ns increment
.TRAN  1NS  80NS
* Plot the results of transient analysis: voltages at nodes 12 and
+13.
.PLOT TRAN V(12) V(13) V(1)
.PROBE
.END
```

The results of the transient analysis are shown in Figure 8.29.

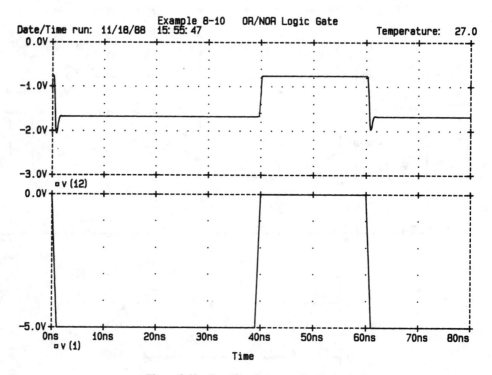

Figure 8.29 Transient response for Example 8.10

SUMMARY

The statements for BJTS are:

```
Q<name>  NC   NB   NE   NS   QNAME   [(area) value]
.MODEL   QNAME   NPN  (P1=V1  P2=V2  P3=V3 ........PN=VN)
.MODEL   QNAME   PNP  (P1=V1  P2=V2  P3=V3 ........PN=VN)
```

REFERENCES

1. H. K. Gummel and H. C. Poon, "An integral charge control model for bipolar transistors," *Bell System Technical Journal,* Vol. 49, January 1970, pp. 827–852.

2. Ian Getreu, *Modeling the bipolar transistor*–Part # 062-2841-00. Beaverton, Ore.: Tektronix Inc., 1979.

3. J. J. Ebers and J. J. Moll, "Large signal behavior of junction transistors," *Proc. IRE,* Vol. 42, December 1954, pp. 1161–1172.

4. L. W. Nagel, *SPICE2—A computer program to simulate semiconductor circuits,* Memorandum no. ERL-M520, May 1975, Electronics Research Laboratory, University of California, Berkeley.

5. A. S. Grove, *Physics and Technology of Semiconductor Devices.* New York: John Wiley, 1967.

6. R. B. Schilling, "A bipolar transistor model for device and circuit design," *RCA Review,* Vol. 32, September 1971, pp. 339–371.

PROBLEMS

8.1. For Example 8.1, calculate the coefficients of Fourier series for the output voltage.

8.2. For Example 8.5, calculate the equivalent input and output noise.

8.3. For Example 8.6, plot the output impedance and the current gain.

8.4. For Figure 8.25, calculate the input and output noise for frequencies from 1 Hz to 10 kHz.

8.5. For Figure 8.25, calculate and plot the frequency response of the output voltage from 10 Hz to 10 MHz in decade steps with 10 points per decade. Assume the peak input voltage is 5 V. The model parameters of the BJTs are BF=50 RB=70 RC=40 TF=0.1NS TR=10NS VJC=0.85 VAF=50. The model parameters of diodes are RS=40 TT=0.1NS.

8.6. For the circuit in Figure P8.6, calculate and plot (a) the magnitude and phase angle of voltage gain, (b) the magnitude of input impedance, and (c) the magnitude of output impedance. The frequency is varied from 1 Hz to 10 MHz in decade steps with 10 points per decade. The peak input voltage is 10 mV. The model parameters of the BJT are IS=2E-16 BF=50 BR=1 RB=5 RC=1 RE=0 CJE=0.4PF VJE=0.8 ME=0.4 CJC=0.5PF VJC=0.8 CCS=1PF VA=100.

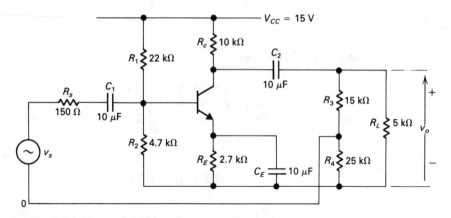

Figure P8.6

8.7. Repeat Problem 8.6 for the circuit in Figure P8.7.

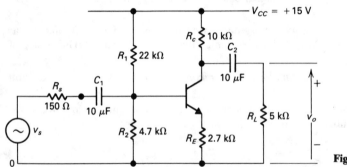

Figure P8.7

8.8. Repeat Problem 8.6 for the circuit in Figure P8.8.

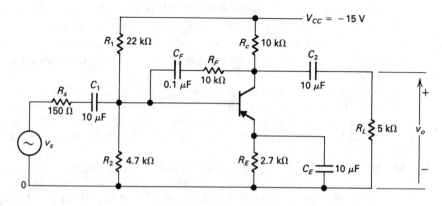

Figure P8.8

8.9. Repeat Problem 8.6 for the circuit in Figure P8.9.

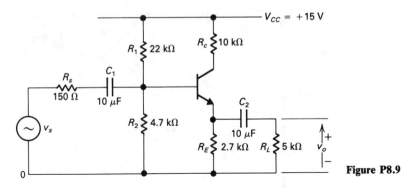

Figure P8.9

8.10. Repeat Problem 8.6 for the circuit in Figure P8.10. Calculate the input and output noise.

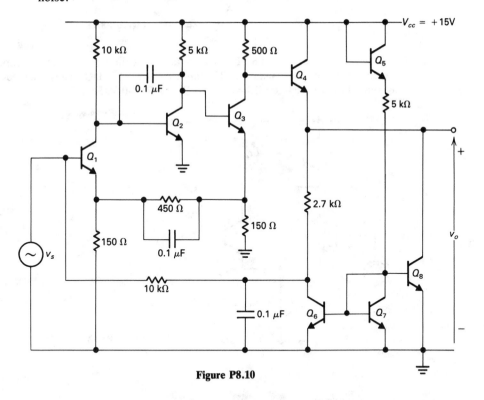

Figure P8.10

8.11. For the circuit in Figure P8.11, calculate and print the DC transfer function (the voltage gain, the input resistance and the output resistance) between the output current and the input voltage V_{EE}. The model parameters of the BJTs are BF=100 BR=1 RB=5 RC=1 RE=0 VJE=0.8 VA=100.

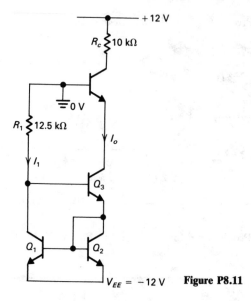

Figure P8.11

8.12. For the circuit in Figure P8.12, calculate and print the voltage gain, the input resistance and the output resistance. The input voltage is 5 V DC. The model parameters of the BJTs are BF=100 BR=1 RB=5 RC=1 RE=0 VJE=0.8 VA=100.

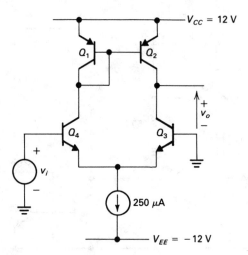

Figure P8.12

9

Field-Effect Transistors

9.1 INTRODUCTION

A field-effect transistor (FET) may be specified by a device statement. *PSpice* generates complex models for FETs. These models are quite complex and incorporate an extensive range of device characteristics, e.g., DC and small-signal behavior, temperature dependency, and noise generation. If such complex models are not necessary, many model parameters can be ignored by the users, and *PSpice* assigns default values to the parameters. The FETs are of three types:

> Junction field-effect transistors
> Metal-oxide silicon field-effect transistors
> Gallium arsenide MESFETs

9.2 JUNCTION FIELD-EFFECT TRANSISTORS

The *PSpice* JFET model is based on the FET model of Schichman and Hodges [1]. The model of an *n*-channel JFET is shown in Figure 9.1. The small-signal model and the static (or DC) model, which are generated by *PSpice,* are shown in

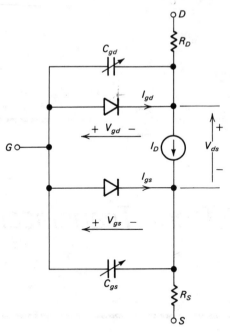

Figure 9.1 *PSpice* n-channel JFET model

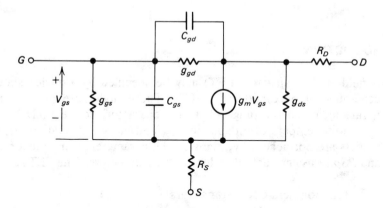

Figure 9.2 Small-signal n-channel JFET model

Figures 9.2 and 9.3, respectively. The model parameters for a JFET device and the default values assigned by *PSpice* are given in Table 9.1. The model equations of JFETs that are used by *PSpice* are described in [1], [3], and [7].

The model statement of an n-channel JFET has the general form

```
.MODEL   JNAME   NJF (P1=V1 P2=V2 P3=V3 ···PN=VN)
```

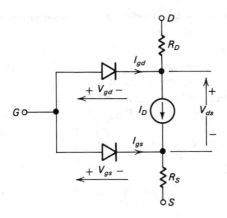

Figure 9.3 Static *n*-channel JFET model

and for a *p*-channel JFET, the statement has the form

```
.MODEL   JNAME   PJF (P1=V1 P2=V2 P3=V3 ···PN=VN)
```

where JNAME is the model name; it can begin with any character and its word size is normally limited to 8. NJF and PJF are the type symbols of *n*-channel and *p*-channel JFETs, respectively. P1, P2, . . . and V1, V2, . . . are the parameters and their values, respectively.

As with diodes and BJTs, an *area factor* is used to determine the number of equivalent parallel JFETs. The model parameters, which are affected by the area

TABLE 9.1 MODEL PARAMETERS OF JFETS

Name	Area	Model parameter	Units	Default	Typical
VTO		Threshold voltage	Volts	−2	−2
BETA	*	Transconductance coefficient	Amps/Volts^2	1E-4 .	1E-3
LAMBDA		Channel-length modulation	Volts^{-1}	0	1E-4
RD	*	Drain ohmic resistance	Ohms	0	100
RS	*	Source ohmic resistance	Ohms	0	100
IS	*	Gate pn saturation current	Amps	1E-14	1E-14
PB		Gate pn potential	Volts	1	0.6
CGD	*	Gate-drain zero-bias pn capacitance	Farads	0	5PF
CGS	*	Gate-source zero-bias pn capacitance	Farads	0	1PF
FC		Forward-bias depletion capacitance coefficient		0.5	
VTOTC		VTO temperature coefficient	Volts/°C	0	
BETATCE		BETA exponential temperature coefficient	percent/°C	0	
KF		Flicker noise coefficient		0	
AF		Flicker noise exponent		1	

factor, are marked by an asterisk (*) in Table 9.1. [(area) value] scales BETA, RD, RS, CGD, CGS, and IS and defaults to 1.

A JFET is modeled as an intrinsic JFET with an ohmic resistance (RD/area) in series with the drain, and another ohmic resistance (RS/area) in series with the source. Positive current is the current which flows into a terminal.

The DC characteristics that are represented by the nonlinear current source I_D are defined (1) by parameters VTO and BETA, which determine the variation of the drain current with the gate voltage, (2) by LAMBDA, which determines the output conductance, and (3) by IS, which determines the reserve saturation current of the two gate junctions. VTO is negative for *n*-channel JFETs and it is positive for *p*-channel JFETs.

The symbol for a JFET is *J*. The name of a JFET must start with *J* and it takes the general form of

```
J〈name〉  ND   NG   NS   JNAME    [(area) value]
```

where ND, NG and NS are the drain, gate and source nodes, respectively.

Some JFET Statements

```
JIM   5   6   8   JNAME
.MODEL   JNAME   NJF
J15   3   9   12   SWITCH   1.5
.MODEL SWITCH   NJF (IS=100E-14 RD=10 RS=10 BETA=1E-3 VTO=-5)
JQ    1   5   9   JMOD
.MODEL JMOD   PJF (IS=100E-14 RD=10 RS=10 BETA=1E-3 CGD=5PF CGS=1PF
+VTO=5)
```

Example 9.1

For the *n*-channel JFET in Figure 9.4, plot the output characteristics if V_{DD} is varied from 0 to 12 V in steps of 0.2 V and V_{GS} is varied from 0 to -4 V in steps of 1 V. The model parameters are IS=100E-14 RD=10 RS=10 BETA=1E-3 VTO=-5.

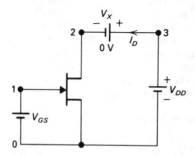

Figure 9.4 A circuit with *n*-channel JFET

Solution The list of the circuit file is as follows.

Example 9.1   Output Characteristics of N-Channel JFET
```
*    Gate to source voltage of 0 V
VGS   1   0   DC   0V
```

```
*    A dummy voltage source of 0 V
VX    3   2   DC   0V
*    DC supply voltage of 12 V
VDD  3   0   DC   12V
*    J1 with model JMOD
J1    2   1   0   JMOD
.MODEL   JMOD   NJF (IS=100E-14 RD=10 RS=10 BETA=1E-3 VTO=-5)
*     VDD is swept from 0 to 12 V and VGS from 0 to -4 V.
.DC   VDD   0   12   0.2   VGS   0   -4   1
.PLOT   DC   I(VX)
.PROBE
.END
```

The output characteristics, which are plots of I_D versus V_{DD}, are shown in Figure 9.5.

Example 9.2

For the JFET in Example 9.1, plot the input characteristic if V_{GS} is varied from 0 to −5 V in steps of 0.1 V and $V_{DD} = 10$ V.

Solution The list of the circuit file is as follows.

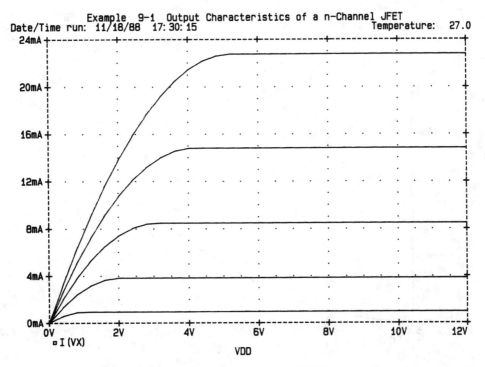

Figure 9.5 Output characteristics of the JFET in Example 9.1

Example 9.2   Input Characteristics of an N-Channel JFET

```
VGS   1   0   DC   0V
VX    3   2   DC   0V
*    DC supply voltage of 10 V
VDD   3   0   DC   10V
*   J1 with model JMOD
J1    2   1   0   JMOD
.MODEL   JMOD   NJF (IS=100E-14 RD=10 RS=10 BETA=1E-3 VTO=-5)
*     VGS is swept from 0 to -5 V.
.DC   VGS   0   -5V   0.1V
.PLOT   DC   I(VX)
.PROBE
.END
```

The input characteristic, which is a plot of I_D versus V_{GS}, is shown in Figure 9.6.

Example 9.3

A JFET transistor amplifier circuit is shown in Figure 9.7. The output is taken from node 6. If the input voltage is $v_{in} = 0.5 \sin(2000\pi t)$, use AC analysis to calculate and print the magnitudes and phase angles of the output voltage, the input current, and the load current. Plot the transient responses of the voltages at nodes 1, 4, and 6

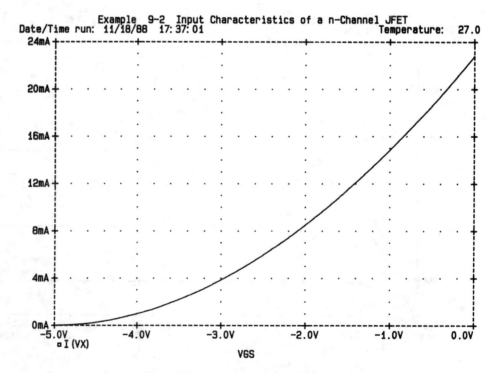

Figure 9.6   Input characteristic for Example 9.2

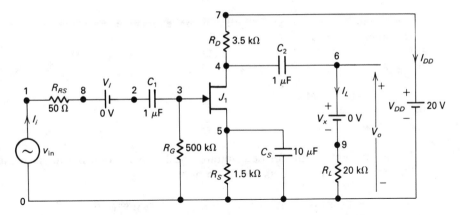

Figure 9.7   N-channel JFET amplifier circuit

from 0 to 1 ms in steps of 10 μs. The model parameters of the JFET are IS=100E-14 RD=10 RS=10 BETA=1E-3 CGD=5PF CGS=1PF VTO=−5. The details of DC analysis and transient analysis operating points should be printed.

Solution   The list of the circuit file is as follows.

Example 9.3   N-Channel JFET Amplifier
```
.OPTIONS  NOPAGE  NOECHO
*  Input voltage has 0.5 V peak at 1 kHz with zero offset value for
*  transient response and 0.5 V peak for frequency response.
VIN  1  0  AC  0.5V  SIN (0  0.5V  1KHZ)
VDD  7  0  DC  20V
*  Dummy voltage source of 0 V
VI  8  2  DC  0V
VX  6  9  DC  0V
RRS  1  8  50
RG  3  0  0.5MEG
RD  7  4  3.5K
RS  5  0  1.5K
RL  9  0  20K
C1  2  3  1UF
C2  4  6  1UF
CS  5  0  10UF
*  N-channel JFET with model JMOD
J1  4  3  5  JMOD
.MODEL JMOD NJF (IS=100E-14 RD=10 RS=10 BETA=1E-3 CGD=5PF CGS=1PF VTO=-5)
*  AC analysis at 1 kHz with a linear increment and only 1 point
.AC LIN  1  1KHZ  1KHZ
*  Transient analysis with details of transient analysis operating point
.TRAN/OP  10US  1MS
*   Print the details of AC analysis operating point.
.OP
* Print the results of AC analysis for the magnitudes of voltages at
```

```
* node 6 and 1 and for the magnitude of current through resistance RRS and
* the current through VX.
.PRINT AC  VM(6) VP(6) IM(RRS) IP(RRS)
.PRINT AC  IM(VI) IP(VI) IM(VX) IP(VX)
*  Plot transient response.
.PLOT TRAN V(6)  V(1)
.PROBE
.END
```

The equivalent circuit for determining the DC bias point is shown in Figure 9.8. The details of the DC bias are given next.

```
****      SMALL-SIGNAL BIAS SOLUTION       TEMPERATURE =   27.000 DEG C
 NODE    VOLTAGE      NODE    VOLTAGE     NODE    VOLTAGE    NODE    VOLTAGE
(    1)    0.0000  (    2)     0.0000  (   3) 8.694E-06  (    4)    11.9300
(    5)    3.4585  (    6)     0.0000  (   7)   20.0000  (    8)     0.0000
(    9)    0.0000
     VOLTAGE SOURCE CURRENTS
     NAME          CURRENT
     VIN          0.000E+00
     VDD         -2.306E-03
     VI           0.000E+00
     VX           0.000E+00
     TOTAL POWER DISSIPATION   4.61E-02  WATTS
```

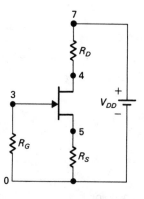

Figure 9.8 Equivalent circuit for DC bias calculation

Once the DC bias points are determined, the small-signal parameters of the JFET in Figure 9.7 are calculated. The details of the operating points are given next.

```
****      OPERATING POINT INFORMATION      TEMPERATURE =   27.000
DEG C
****   JFETS
NAME          J1
MODEL         JMOD
ID            2.31E-03
```

```
VGS              -3.46E+00
VDS               8.47E+00
GM                3.04E-03
GDS               0.00E+00
CGS               4.72E-13
CGD               1.39E-12
```

The outputs at a frequency of 1 kHz are as follows.

```
FREQ        VM(6)         VP(6)         IM(RRS)       IP(RRS)
1.000E+03   4.382E+00    -1.769E+02     9.990E-07     2.555E+00
FREQ        IM(VI)        IP(VI)        IM(VX)        IP(VX)
1.000E+03   9.990E-07     2.555E+00     2.191E-04    -1.769E+02
```

The equivalent circuit for determining the transient analysis bias point is shown in Figure 9.9. The transient analysis bias point and the operating point are the same as that of the DC analysis because the capacitors do not have any initial voltages. The transient response are shown in Figure 9.10.

Example 9.4

If the JFET in Figure 9.7 is replaced by the subcircuit model of Figure 9.11, plot the frequency response of the output voltage. The frequency is varied from 10 Hz to 100 MHz with a decade increment and 10 points per decade.

Solution   The list of the circuit file is as follows.

Example 9.4   N-Channel JFET Amplifier
```
.OPTIONS  NOPAGE  NOECHO
*  Input voltage has 0.5 V peak for frequency response.
VIN   1   0   AC   0.5V
VDD   7   0   DC   20V
```

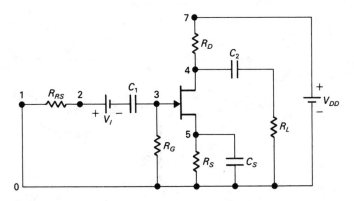

Figure 9.9 Equivalent circuit for transient analysis bias calculation

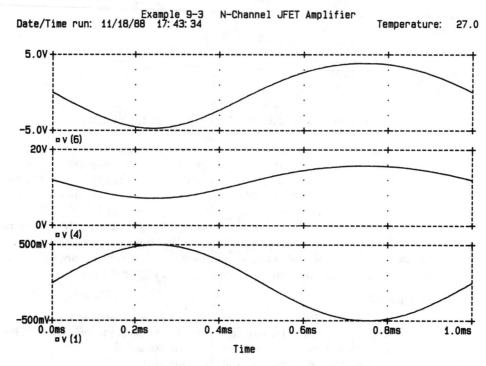

Figure 9.10 Transient responses for Example 9.3

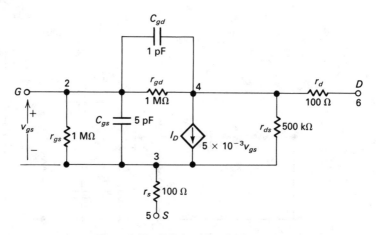

Figure 9.11 Subcircuit model for JFET

```
*    Dummy voltage source of OV
VI    8   2   DC   OV
VX    6   9   DC   OV
RRS   1   8   50
RG    3   0   0.5MEG
RD    7   4   3.5K
RS    5   0   1.5K
RL    9   0   20K
C1    2   3   1UF
C2    4   6   1UF
CS    5   0   100UF
*    Calling subcircuit for TRANS
XQ1   4   3   5   TRANS
*    Subcircuit definition for TRANS
.SUBCKT   TRANS   6   2   5
RD    4   6   100
RS    3   5   100
RGS   2   3   1MEG
CGS   2   3   5PF
RGD   2   4   1MEG
CGS   2   4   1PF
RDS   4   3   500K
*    Voltage-controlled current source with a gain of 5E-3
G1    4   3   2   3   5E-3
.ENDS   TRANS
*    AC analysis for 100 Hz to 100 MHz with a decade increment and
*    10 points per decade
.AC   DEC   10   10HZ   100MEGHZ
*    Plot the results of ac analysis for the magnitudes and phases of output
*    voltage and the magnitudes of input and load currents
.PLOT   AC   VM(6)   VP(6)
.PLOT   AC   IM(VI)   IM(VX)
.PROBE
.END
```

The frequency response for example 9.4 is shown in Figure 9.12.

Example 9.5

A *p*-channel JFET bootstrapped amplifier is shown in Figure 9.13. The output is taken from node 5. Calculate and print the voltage gain, the input resistance, and the output resistance. The model parameters of the JFET are IS=100E-14 RD=10 RS=10 BETA=1E-3 VTO=5.

Solution The list of the circuit is as follows.

```
Example 9.5     Bootstrapped JFET Amplifier
VDD   0   6   15V
*    Input voltage of 5 V DC
```

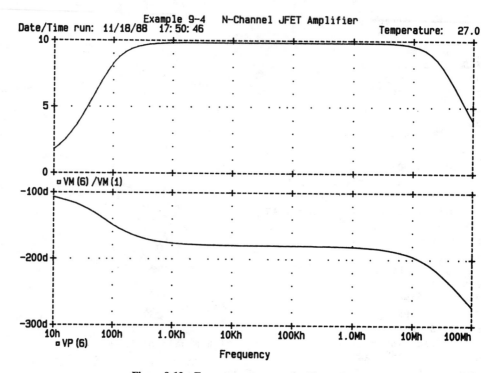

Figure 9.12   Frequency response for Example 9.4

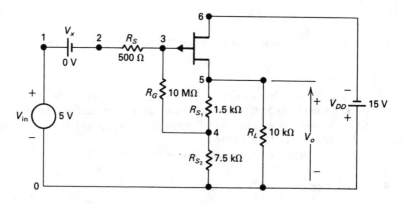

Figure 9.13   P-channel JFET bootstrapped amplifier

```
VIN   1   0   DC   5V
*   A dummy voltage source of 0 V
VX    1   2   DC   0V
RS    2   3   500
RG    3   4   10MEG
RS1   5   4   1.5K
RS2   4   0   7.5K
RL    5   0   10K
*   P-channel JFET of model JMOD
JX    6   3   5   JMOD
*   Model statement for P-channel JFET
.MODEL JMOD PJF (IS=100E-14 RD=10 RS=10 BETA=1E-3 VTO=5)
*   Transfer function analysis between the output and input voltages
.TF  V(5)  VIN
.END
```

The results of the transfer function analysis are

```
****      SMALL-SIGNAL CHARACTERISTICS
          V(5)/VIN = -9.257E-01
          INPUT RESISTANCE AT VIN =  4.375E+07
          OUTPUT RESISTANCE AT V(5) =  3.521E+02
          JOB CONCLUDED
          TOTAL JOB TIME              9.50
```

9.3 METAL OXIDE SILICON FIELD-EFFECT TRANSISTORS

The *PSpice* model of an N-channel MOSFET [1–3] is shown in Figure 9.14. The small-signal model and the static (or DC) model generated by *PSpice* are shown in Figures 9.15 and 9.16, respectively. The model parameters for a MOSFET device and the default values assigned by *PSpice* are given in Table 9.2. The model equations of MOSFETs that are used by *PSpice* are described in [1], [3], and [7].

The model statement of N-channel MOSEFTs has the general form

```
    .MODEL  MNAME  NMOS (P1=V1  P2=V2  P3=V3 ··· PN=VN)
```

and the statement for P-channel MOSFETs has the form

```
    .MODEL  MNAME  PMOS (P1=V1  P2=V2  P3=V3 ··· PN=VN)
```

where MNAME is the model name; it can begin with any character, and its word size is normally limited to 8. NMOS and PMOS are the type symbols of N-channel and P-channel MOSFETs, respectively. P1, P2, . . . and V1, V2, . . . are the parameters and their values, respectively.

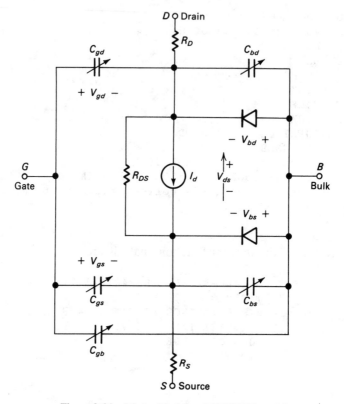

Figure 9.14 *PSpice* N-channel MOSFET model

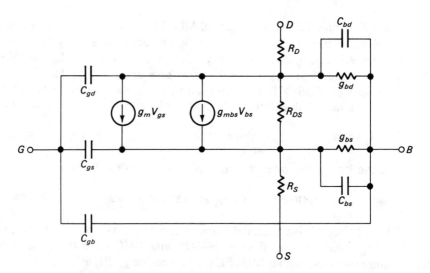

Figure 9.15 Small-signal N-channel MOSFET model

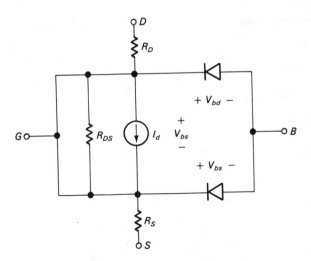

Figure 9.16   Static N-channel MOSFET model

TABLE 9.2   MODEL PARAMETERS OF MOSFETS

Name	Model parameters	Units	Default	Typical
LEVEL	Model type (1, 2, or 3)		1	
L	Channel length	meters	DEFL	
W	Channel width	meters	DEFW	
LD	Lateral diffusion length	meters	0	
WD	Lateral diffusion width	meters	0	
VTO	Zero-bias threshold voltage	Volts	0	0.1
KP	Transconductance	Amps/Volts^2	2E-5	2.5E-5
GAMMA	Bulk threshold parameter	Volts$^{1/2}$	0	0.35
PHI	Surface potential	Volts	0.6	0.65
LAMBDA	Channel-length modulation (LEVEL =1 or 2)	Volts^{-1}	0	0.02
RD	Drain ohmic resistance	Ohms	0	10
RS	Source ohmic resistance	Ohms	0	10
RG	Gate ohmic resistance	Ohms	0	1
RB	Bulk ohmic resistance	Ohms	0	1
RDS	Drain-source shunt resistance	Ohms	∞	
RSH	Drain-source diffusion sheet resistance	Ohms/square	0	20
IS	Bulk pn saturation current	Amps	1E-14	1E-15
JS	Bulk pn saturation current/area	Amps/meters^2	0	1E-8
PB	Bulk pn potential	Volts	0.8	0.75
CBD	Bulk-drain zero-bias *pn* capacitance	Farads	0	5PF
CBS	Bulk-source zero-bias *pn* capacitance	Farads	0	2PF
CJ	Bulk pn zero-bias bottom capacitance/length	Farads/meter^2	0	
CJSW	Bulk pn zero-bias perimeter capacitance/length	Farads/meters	0	

TABLE 9.2 CONTINUED

Name	Model parameters	Units	Default	Typical
MJ	Bulk pn bottom grading coefficient		0.5	
MJSW	Bulk pn sidewall grading coefficient		0.33	
FC	Bulk pn forward-bias capacitance coefficient		0.5	
CGSO	Gate-source overlap capacitance/ channel width	Farads/meters	0	
CGDO	Gate-drain overlap capacitance/ channel width	Farads/meters	0	
CGBO	Gate-bulk overlap capacitance/ channel length	Farads/meters	0	
NSUB	Substrate doping density	1/centimeter^3	0	
NSS	Surface state density	1/centimeter^2	0	
NFS	Fast surface state density	1/centimeter^2	0	
TOX	Oxide thickness	meters	∞	
TPG	Gate material type: +1 = opposite of substrate, −1 = same as substrate, 0 = aluminum		+1	
XJ	Metallurgical junction depth	meters	0	
UO	Surface mobility	centimeters^2/Volts·seconds	600	
UCRIT	Mobility degradation critical field (LEVEL = 2)	Volts/centimeter	1E4	
UEXP	Mobility degradation exponent (LEVEL = 2)		0	
UTRA	(Not used) mobility degradation transverse field coefficient			
VMAX	Maximum drift velocity	meters/second	0	
NEFF	Channel charge coefficient (LEVEL = 2)		1	
XQC	Fraction of channel charge attributed to drain		1	
DELTA	Width effect on threshold		0	
THETA	Mobility modulation (LEVEL = 3)	Volts^{-1}	0	
ETA	Static feedback (LEVEL = 3)		0	
KAPPA	Saturation field factor (LEVEl = 3)		0.2	
KF	Flicker noise coefficient		0	1E-26
AF	Flicker noise exponent		1	1.2

L and W are the channel length and width, respectively. AD and AS are the drain and source diffusion areas. L is decreased by twice LD to get the effective channel length. W is decreased by twice WD to get the effective channel width. L and W can be specified on the device, the model, or on the .OPTION statement. The value on the device supersedes the value on the model, which supersedes the value on the .OPTION statement.

AD and AS are the drain and source diffusion areas. PD and PS are the drain and source diffusion perimeters. The drain-bulk and source-bulk saturation currents can be specified either by JS, which is multiplied by AD and AS, or by IS, which is an absolute value. The zero-bias depletion capacitances can be specified by CJ, which is multiplied by AD and AS, and by CJSW, which is multiplied by PD and PS. Alternately, these capacitances can be set by CBD and CBS, which are absolute values.

A MOSFET is modeled as an intrinsic MOSFET with ohmic resistances in series with the drain, source, gate, and bulk (substrate). There is also a shunt resistance (RDS) in parallel with the drain-source channel. NRD, NRS, NRG, and NRB are the relatively resistivities of the drain, source, gate, and substrate in squares. These parasitic (ohmic) resistances can be specified either by RSH, which is multiplied by NRD, NRS, NRG, and NRB, respectively. Alternatively, the absolute values of RD, RS, RG, and RB can be specified directly.

PD and PS default to 0. NRD and NRS default to 1. NRG and NRB default to 0. Defaults for L, W, AD, and AS may be set in the .OPTIONS statement. If AD or AS defaults are not set, they also default to 0. If L or W defaults are not set, they default to 100 μm.

The dc characteristics are defined by parameters VTO, KP, LAMDA, PHI, and GAMMA, which are computed by *PSpice* by using the fabrication-process parameters NSUB, TOX, NSS, NFS, TPG, etc. The values of VTO, KP, LAMDA, PHI, and GAMMA, which are specified on the model statement, superseded the values calculated by *PSpice* based on fabrication-process parameters. *VTO is positive for enhancement type N-channel MOSFET and for depletion type p-channel MOSFET. VTO is negative for enhancement type P-channel MOSFET and for depletion type n-channel MOSFET.*

PSpice incorporates three MOSFET device models. The LEVEL parameter selects among different models for the intrinsic MOSFET. If LEVEL = 1, the Schichman-Hodges model [1] is used. If LEVEL = 2, an advanced version of the Schichman-Hodges model, which is a geometry-based analytical model and incorporates extensive second-order effects [3], is used. If LEVEL = 3, a modified version of the Schichman-Hodges model, which is a semiempirical short-channel model [3], is used.

The LEVEL 1 model, which employs fewer fitting parameters, gives approximate results. However, it is useful for a quick and rough estimate of the circuit performances and it is normally adequate for the analysis of basic electronic circuits. The LEVEL 2 model, which can take into considerations of various parameters, requires a great amount of CPU time for the calculations and could cause convergence problem. The LEVEL 3 model introduces a smaller error as compared to that of LEVEL 2 model and the CPU time is also approximately 25% less. The LEVEL 3 model is designed for MOSFETs with short channel.

The symbol for a metal-oxide silicon field-effect transistor (MOSFET) is *M*. The name of MOSFETs must start with *M* and takes the general form of

```
M⟨name⟩   ND   NG   NS   NB   MNAME
+         [L=⟨value⟩] [W=⟨value⟩]
+         [AD=⟨value⟩] [AS=⟨value⟩]
+         [PD=⟨value⟩] [PS=⟨value⟩]
+         [NRD=⟨value⟩] [NRS=⟨value⟩]
+         [NRG=⟨value⟩] [NRB=⟨value⟩]
```

where ND, NG, NS, and NB are the drain, gate, source, and bulk (or substrate) nodes, respectively. MNAME is the model name, and it can begin with any character; its word size is normally limited to 8. Positive current is the current that flows into a terminal. That is, the current flows from the drain node through the device to the source node for an *n*-channel MOSFET.

Some MOSFET Statements

```
M1    4   2   7    0 MMOD   L=10U W=20U
.MODEL   MMOD   NMOS
M13   15   3   0    0 IRF150
.MODEL   IRF150   NMOS (LEVEL=3 TOX=.10U L=3.0U LD=.5U W=2.0 WD=0
+XJ=1.2U
+        NSUB=4E14 IS=2.1E-14 RB=0 RD=.01 RS=.03 RDS=1E6 VTO=3.25
+        UO=550 THETA=.1 ETA=0 VMAX=1E6 CBS=1P CBD=4000P PB=.7 MJ=.5
+        RG=4.9 CGSO=1690P CGDO=365P CGBO=1P)
M2A   0   2   20   20 IRF9130
.MODEL   IRF9130   PMOS (LEVEL=3 TOX=.1U L=3.0U LD=.5U W=1.3 WD=0
+XJ=1.2U
+        NSUB=4E14 IS=2.1E-14 RB=0 RD=.03 RS=.2 RDS=5E5 VTO=-3.7
+        UO=600 THETA=.1 ETA=0 VMAX=1E6 CBS=1P CBD=2000P PB=.7 MJ=.5
+        RG=5 CGSO=520P CGDO=180P CGBO=1P)
MA    0   2   15   15 PMOD   L=20U W=20U AD=100U AS=200U PD=50U
+                             PS=50U NRD=10 NRS=20 NRG=10
.MODEL   PMOD   PMOS
```

Example 9.6

An N-channel enhancement-type MOSFET amplifier with series-shunt feedback is shown in Figure 9.17. Plot the magnitude of output voltage. The frequency is varied from 10 Hz to 100 MHz in decade steps with 10 points per decade. The peak input voltage is 100 mV. The model parameters of the MOSFET are VTO=1 KP=6.5E-3 CBD=5PF CBS=2PF RD=5 RS=2 RB=0 RG=0 RDS=1MEG CGSO=1PF CGDO=1PF CGBO=1PF. Print the details of the bias and operating points. **Solution**   The list of the circuit file is as follows.

Example 9.6   An MOSFET Feedback Amplifier
```
*   Input voltage of 100 mV peak for frequency response
VIN   1   7   AC   100mV
VDD   8   0   15V
```

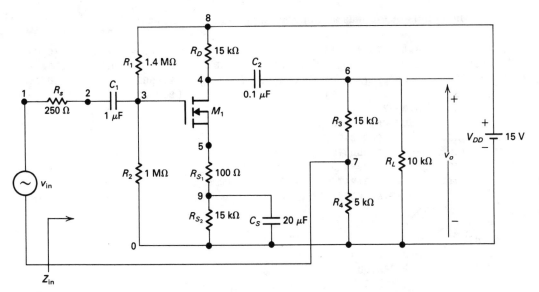

Figure 9.17   A MOSFET feedback amplifier

```
RS    1  2  250
C1    2  3  1UF
R1    8  3  1.4MEG
R2    3  0  1MEG
RD    8  4  15K
RS1   5  9  100
RS2   9  0  15K
CS    9  0  20UF
C2    4  6  0.1UF
R3    6  7  15K
R4    7  0  5K
RL    6  0  10K
*     MOSFET M1 with model MQ is connected to 4 (drain), 3 (gate), 5 (source)
*     and 5 (substrate).
M1    4  3  5  5  MQ
*     Model for MQ
.MODEL MQ NMOS (VTO=1 KP=6.5E-3 CBD=5PF CBS=2PF RD=5 RS=2 RB=0
+     RG=0 RDS=1MEG CGSO=1PF CGDO=1PF CGBO=1PF)
*     AC analysis for 10 Hz to 100 MHz with a decade increment and 10
*     points per decade
.AC   DEC 10  10HZ  100MEGHZ
*     Plot the results of ac analysis: voltage at node 6.
.PLOT AC  VM(6)
*     Print the details of DC operating point.
.OP
.PROBE
.END
```

The details of the bias and operating points are given next.

```
****      SMALL-SIGNAL BIAS SOLUTION      TEMPERATURE =  27.000 DEG C
 NODE   VOLTAGE      NODE   VOLTAGE      NODE   VOLTAGE     NODE   VOLTAGE
(   1)   0.0000  (    2)   0.0000  (    3)   6.2500  (    4)  10.1000
(   5)   4.9323  (    6)   0.0000  (    7)   0.0000  (    8)  15.0000
(   9)   4.8997
        VOLTAGE SOURCE CURRENTS
        NAME          CURRENT
        VIN           0.000E+00
        VDD          -3.329E-04
        TOTAL POWER DISSIPATION   4.99E-03   WATTS
****      OPERATING POINT INFORMATION      TEMPERATURE =  27.000 DEG C
**** MOSFETS
NAME         M1
MODEL        MQ
ID           3.32E-04
VGS          1.32E+00
VDS          5.17E+00
VBS          0.00E+00
VTH          1.00E+00
VDSAT        3.17E-01
GM           2.06E-03
GDS          1.00E-06
GMB          0.00E+00
CBD          1.83E-12
CBS          2.00E-12
CGSOV        1.00E-16
CGDOV        1.00E-16
CGBOV        1.00E-16
CGS          0.00E+00
CGD          0.00E+00
CGB          0.00E+00
        JOB CONCLUDED
        TOTAL JOB TIME          17.41
```

The frequency response for Example 9.6 is shown in Figure 9.18.

Example 9.7

For Figure 9.17, plot the magnitude response of output impedance. The frequency is varied from 10 Hz to 100 MHz in decade steps with 10 points per decade.

Solution The output impedance of the MOSFET feedback amplifier in Figure 9.17 can be determined by short-circuiting the input source and connecting a test current source between terminals 0 and 6, as shown in Figure 9.19. Let the peak value of the test current be 1 mA. The voltage at node 6 is a measure of the output impedance: $Z_{out} = V(6)/1 \text{ mA} = V(6) \text{ k}\Omega$.

The list of the circuit file is as follows.

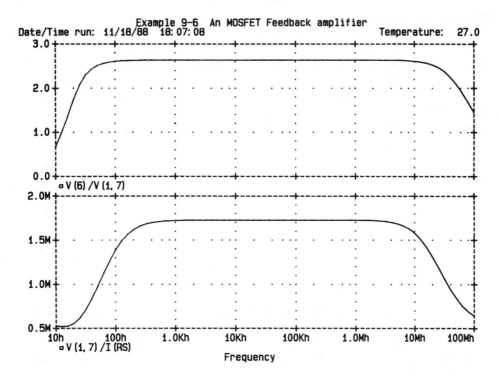

Figure 9.18   Frequency response for Example 9.6

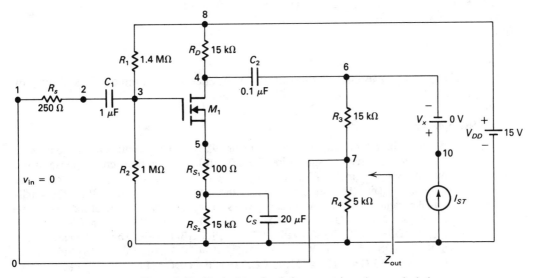

Figure 9.19   Equivalent circuit for output impedance calculation

Example 9.7   Output Impedance of a MOSFET Feedback amplifier

```
*    Input source VIN is shorted.
VIN   1   9   AC   0V
*    Test current of 1 mA peak for frequency response
IST   0   10   AC   1MA
*    A dummy source of 0 V DC
VX    10   6   DC 0V
VDD   8   0   15V
RS    1   2   250
C1    2   3   1UF
R1    8   3   1.4MEG
R2    3   0   1MEG
RD    8   4   15K
RS1   5   9   100
RS2   9   0   15K
CS    9   0   20UF
C2    4   6   0.1UF
R3    6   7   15K
R4    7   0   5K
RL    6   0   10K
*    M1 with model MQ, whose substrate is connected to node 5
M1 4  3  5  5  MQ
*    Model for n-channel MOSFET with model name MQ
.MODEL MQ NMOS (VTO=1 KP=6.5E-3 CBD=5PF CBS=2PF RD=5 RS=2 RB=0
+    RG=0 RDS=1MEG CGSO=1PF CGDO=1PF CGBO=1PF)
*    AC analysis for 10 Hz to 100 MHz with a decade increment and 10
*    points per decade
.AC   DEC   10   10HZ   10MEGHZ
*    Plot the results of AC analysis: voltage at node 6.
.PLOT   AC   VM(6)
.PROBE
.END
```

The frequency response of the output impedance for Example 9.7 is shown in Figure 9.20.

Example 9.8

A CMOS inverter circuit is shown in Figure 9.21(a). The output is taken from node 3. The input voltage is shown in Figure 9.21(b). Plot the transient response of the output voltage from 0 to 80 μs in steps of 2 μs. If the input voltage is 5 V, calculate the voltage gain, the output resistance, and the output resistance. Print the small-signal parameters of the PMOS and NMOS. The model parameters of the PMOS are L=1U W=20U VTO=-2 KP=4.5E-4 CBD=5PF CBS=2PF RD=5 RS=2 RB=0 RG=0 RDS=1MEG CGSO=1PF CGDO=1PF CGBO=1PF. The model parameters of the NMOS are L=1U W=5U VTO=2 KP=4.5E-5 CBD=5PF CBS=2PF RD=5 RS=2 RB=0 RG=0 RDS=1MEG CGSO=1PF CGDO=1PF CGBO=1PF.

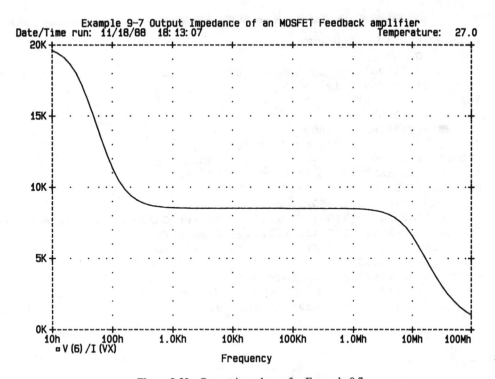

Figure 9.20 Output impedance for Example 9.7

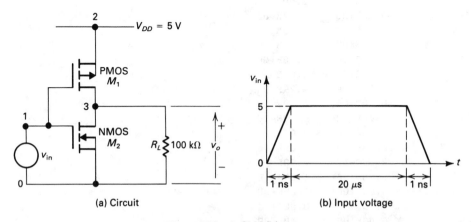

Figure 9.21 A CMOS inverter

Solution   The list of the circuit file is as follows.

Example 9.8   A CMOS Inverter

```
VDD   2   0   5V
*  The input voltage is 5 V for DC analysis and pulse waveform for
*  transient analysis.
VIN   1   0   DC  5V  PULSE (0  5V  0  1NS  1NS  20US  40US)
RL    3   0   100K
*    PMOS with model PMOD
M1    3   1   2   2   PMOD  L=1U  W=20U
.MODEL PMOD PMOS (VTO=-2 KP=4.5E-4 CBD=5PF CBS=2PF RD=5 RS=2 RB=0
+  RG=0 RDS=1MEG CGSO=1PF CGDO=1PF CGBO=1PF)
M2    3   1   0   0   NMOD  L=1U  W=5U
*    NMOS with model NMOD
.MODEL NMOD NMOS (VTO=2 KP=4.5E-5 CBD=5PF CBS=2PF RD=5 RS=2 RB=0
+  RG=0 RDS=1MEG CGSO=1PF CGDO=1PF CGBO=1PF)
*  Transient analysis from 0 to 80μs in steps of 1 μs
.TRAN  1US  80US
*  Transfer function analysis
.TF  V(3)  VIN
*  Print details of operating points.
.OP
.PLOT TRAN  V(3)  V(1)
.PROBE
.END
```

```
    ****     OPERATING POINT INFORMATION        TEMPERATURE =   27.000 DEG C
**** MOSFETS
NAME         M1          M2
MODEL        PMOD        NMOD
ID          -5.00E-06    1.53E-11
VGS          0.00E+00    5.00E+00
VDS         -5.00E+00    2.27E-08
VBS          0.00E+00    0.00E+00
VTH         -2.00E+00    2.00E+00
VDSAT        0.00E+00    3.00E+00
GM           0.00E+00    5.09E-12
GDS          1.00E-06    6.76E-04
GMB          0.00E+00    0.00E+00
CBD          1.86E-12    5.00E-12
CBS          2.00E-12    2.00E-12
CGSOV        2.00E-17    5.00E-18
CGDOV        2.00E-17    5.00E-18
CGBOV        1.00E-18    1.00E-18
CGS          0.00E+00    0.00E+00
CGD          0.00E+00    0.00E+00
CGB          0.00E+00    0.00E+00
```

```
****     SMALL-SIGNAL CHARACTERISTICS
   V(3)/VIN = -7.422E-09
   INPUT RESISTANCE AT VIN =   1.000E+20
   OUTPUT RESISTANCE AT V(3) =   1.462E+03
      JOB CONCLUDED
      TOTAL JOB TIME           48.33
```

The frequency response for Example 9.8 is shown in Figure 9.22.

9.4 GALLIUM ARSENIDE MESFETs

The *PSpice* model of an N-channel GaAsFET (gallium arsenide FET) is shown in Figure 9.23 [5, 6]. The small-signal model, which is generated by *PSpice,* is shown in Figure 9.24. The model parameters for a GaAsFET device and the default values assigned by *PSpice* are listed in Table 9.3. The model equations of GaAsFETs that are used by *PSpice* are described in [5], [6], and [7].

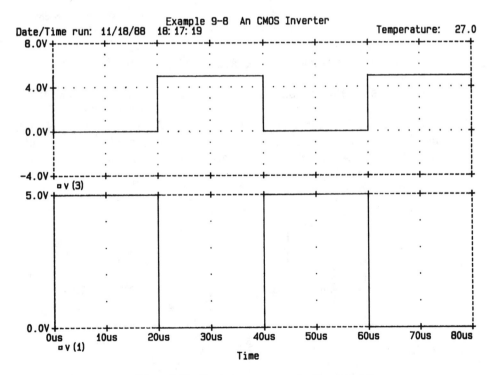

Figure 9.22   Frequency response for Example 9.8

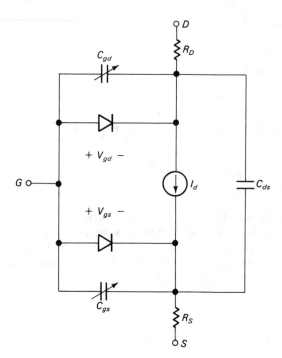

Figure 9.23 *PSpice* N-channel
GaAsFET Model

The model statement of N-channel GaAsFETs has the general form

```
.MODEL  BNAME  GASFET (P1=V1  P2=V2  P3=V3···PN=VN)
```

where GASFET is the type symbol of *n*-channel GaAsFETs. BNAME is the
model name. It can begin with any character and its word size is normally limited
to 8. P1, P2, . . . and V1, V2, . . . are the parameters and their values, respectively.

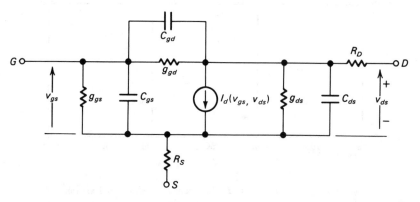

Figure 9.24 Small-signal N-channel GaAsFET model

TABLE 9.3 MODEL PARAMETERS OF GaAs MESFETs

Name	Area	Model parameters	Units	Default	Typical
VTO		Threshold voltage	Volts	−2.5	−2.0
ALPHA		tanh constant	Volts^{-1}	2.0	1.5
BETA		Transconductance coefficient	Amps/Volts^2	0.1	25U
LAMBDA		Channel-length modulation	Volts	0	1E-10
RG	*	Gate ohmic resistance	Ohms	0	1
RD	*	Drain ohmic resistance	Ohms	0	1
RS	*	Source ohmic resistance	Ohms	0	1
IS		Gate pn saturation current	Amps	1E-14	
M		Gate pn grading coefficient		0.5	
N		Gate pn emission coefficient		1	
VBI		Threshold voltage	Volts	1	0.5
CGD		Gate-drain zero-bias pn capacitance	Farads	0	1FF
CGS		Gate-source zero-bias pn capacitance	Farads	0	6FF
CDS		Drain-source capacitance	Farads	0	0.3FF
TAU		Transit time	seconds	0	10PS
FC		Forward-bias depletion capacitance coefficient		0.5	
VTOTC		VTO temperature coefficient	Volts/°C	0	
BETATCE		BETA exponent temperature coefficient	%/°C	0	
KF		Flicker noise coefficient		0	
AF		Flicker noise exponent		1	

The GaAsFET is modeled as an intrinsic FET with an ohmic resistance (RD/area) in series with the drain, another ohmic resistance (RS/area) in series with the source, and with another ohmic resistance (RG/area) in series with the gate. [(area) value] is the relative device area and defaults to 1.0.

The symbol for a gallium arsenide MESFET (GaAs MESFET or GaAsFET) is *B*. The name of a GaAs MESFET must start with *B* and it takes the general form

```
B⟨name⟩ ND  NG  NS  BNAME  [(area) value]
```

where ND, NG, ND are the drain, gate, and source nodes, respectively. BNAME, which is the model name, can begin with any character, and its word size is normally limited to 8. Positive current flows into a terminal.

Some GaAs MESFET Statements

```
BIX  2  5  7  NMOD
.MODEL NMOD GASFET
BIM 15  1  0  GMOD
.MODEL GMOD GASFET (VTO=-2.5 BETA=60U VBI=0.5 ALPHA=1.5 TAU=10PS)
B5  7  9  3   MNOM 1.5
.MODEL MNOM GASFET (VTO=-2.5 BETA=32U VBI=0.5 ALPHA=1.5)
```

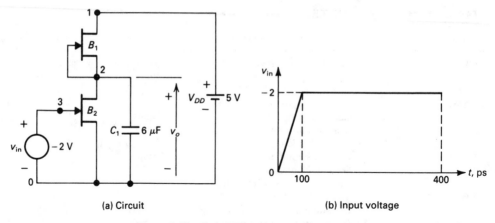

(a) Circuit   (b) Input voltage

Figure 9.25   GaAsFET inverting with active load

Example 9.9

A GaAsFET inverter with active load is shown in Figure 9.25(a). The input voltage is a pulse waveform, as shown in Figure 9.25(b). Plot the transient response of the output voltage for a time duration of 240 ps in steps of 2 ps. Plot the DC transfer characteristic if the input voltage is varied from −2.5 V to 1 V in steps of 0.1 V. The model parameters of the GaAsFET are VTO=−2 BETA=60U VBI=0.5 ALPHA=1.5 TAU=10PS and those of B2 are VTO=−2.BETA=3U VBI=0.5 ALPHA=1.5. Calculate the DC voltage gain, the input resistance, and the output resistance. Print the small-signal parameters for DC analysis.

Solution   The list of the circuit file is as follows.

```
Example 9.9     A GaAsFET inverter with active load
VDD  1  0  5V
*    Pulsed input voltage
VIN  3  0  DC  -2V  PWL (0  0  100PS -2V  1NS -2V)
*    GaAsFET, which is connected to 1 (drain), 2 (gate) and
*    2 (source), has a model of GF1.
B1   1  2  2  GF1
B2   2  4  0  GF2
C1   2  0  6F  IC=0V
RS   3  4  50
*    Model for GF1
.MODEL GF1 GASFET (VTO=-2.5 BETA=65U VBI=0.5 ALPHA=1.5 TAU=10PS)
.MODEL GF2 GASFET (VTO=-2.5 BETA=32.5U VBI=0.5 ALPHA=1.5)
*    Transient analysis for 0 to 240 ps with 2-ps increment
.TRAN  2PS  240PS  UIC
.DC VIN  -2.5  1  0.1
*    Plot the results of transient analysis.
.PLOT TRAN  V(3)  V(2)
.PLOT  DC  V(2)
```

```
*  DC transfer characteristics
.TF  V(2)  VIN
*   Small-signal parameters for DC analysis
.OP
.PROBE
.END
```

```
****     SMALL-SIGNAL BIAS SOLUTION      TEMPERATURE =   27.000 DEG C
NODE    VOLTAGE      NODE   VOLTAGE      NODE   VOLTAGE      NODE   VOLTAGE
(   1)    5.0000  (    2)    4.8787  (    3)   -1.0000  (    4)   -1.0000
       VOLTAGE SOURCE CURRENTS
       NAME           CURRENT
       VDD          -7.313E-05
       VIN           6.899E-12
       TOTAL POWER DISSIPATION   3.66E-04   WATTS


****     OPERATING POINT INFORMATION     TEMPERATURE =   27.000 DEG C
**** GASFETS
NAME         B1          B2
MODEL        GF1         GF2
ID           7.31E-05    7.31E-05
```

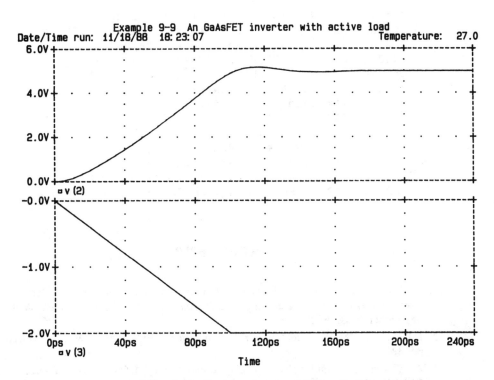

Figure 9.26 Transient response for Example 9.9

```
VGS         0.00E+00    -1.00E+00
VDS         1.21E-01     4.88E+00
GM          5.85E-05     9.75E-05
GDS         5.90E-04     1.93E-10
CGS         0.00E+00     0.00E+00
CGD         0.00E+00     0.00E+00
CDS         0.00E+00     0.00E+00


****     SMALL-SIGNAL CHARACTERISTICS
         V(2)/VIN = -1.654E-01
         INPUT RESISTANCE AT VIN =   4.618E+11
         OUTPUT RESISTANCE AT V(2) =  1.696E+03
```

The transient response and the DC transfer characteristics for Example 9.9 is shown in Figure 9.26.

SUMMARY

The model statements for FETs can be summarized as

```
B(name) ND  NG  NS  BNAME   [(area) value]
.MODEL  BNAME   GASFET (P1=V1 P2=V2 P3=V3 ...PN=VN)
J(name) ND  NG  NS  JNAME [(area) value]
.MODEL  JNAME   NJF (P1=V1 P2=V2 P3=V3 ...PN=VN)
.MODEL  JNAME   PJF (P1=V1 P2=V2 P3=V3 ...PN=VN)
M(name) ND  NG  NS  NB  MNAME
+       [L=(value)] [W=(value)]
+       [AD=(value)] [AS=(value)]
+       [PD=(value)] [PS=(value)]
+       [NRD=(value)] [NRS=(value)]
+       [NRG=(value)] [NRB=(value)]
.MODEL  MNAME   NMOS (P1=V1 P2=V2 P3=V3 ...PN=VN)
.MODEL  MNAME   PMOS (P1=V1 P2=V2 P3=V3 ...PN=VN)
```

REFERENCES

1. H. Schichman and D. A. Hodges, "Modeling and simulation of insulated gate field effect transistor switching circuits," *IEEE Journal of Solid-State Circuits,* Vol. SC-3, September 1968, pp 285–289.

2. J. F. Meyer, "MOS models and circuit simulation," *RCA Review,* Vol. 32, March 1971, pp. 42–63.

3. A. Vladimirescu and Sally Liu, *The simulation of MOS integrated circuits using SPICE2,* Memorandum no. M80/7, February 1980, University of California, Berkeley.

4. L. M. Dang, "A simple current model for short channel IGFET and its application to circuit simulation," *IEEE Journal of Solid-State Circuits,* Vol. SC-14, No. 2, 1979, pp. 358–367.

5. W. R. Curtice, "A MESFET model for use in the design of GaAs integrated circuits," *IEEE Transactions on Microwave Theory and Techniques,* Vol. MTT-23, May 1980, pp 448–456.

6. S. E. Sussman-Fort, S. Narasimhan, and K. Mayaram, "A complete GaAs MESFET computer model for SPICE," *IEEE Transactions on Microwave Theory and Techniques,* Vol. MTT-32, April 1984, pp 471–473.

7. *PSpice Manual.* Irvine, Calif.: MicroSim Corporation, 1988.

PROBLEMS

9.1. For the amplifier circuit in Figure 6.7, calculate and plot the frequency responses of the output voltage and the input current. The frequency is varied from 10 Hz to 10 MHz in decade steps with 10 points per decade.

9.2. A shunt-shunt feedback is applied to the amplifier circuit in Figure 9.7. This is shown in Figure P9.2. Calculate and print the frequency responses of the output voltage and the input current. The frequency is varied from 10 Hz to 10 MHz in steps of decade and 10 points per decade.

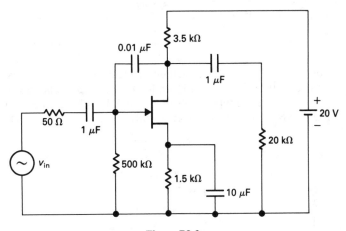

Figure P9.2

9.3. Repeat Example 9.5 if the transistor is an N-channel JFET. The model parameters are IS=100E−14 RD=10 RS=10 BETA=1E−3 VTO=−2. Assume $R_{S2} = 0$.

9.4. For the N-channel enhancement-type MOSFET in Figure P9.4, plot the output characteristics if V_{DS} is varied from 0 to 15 V in steps of 0.1 V and V_{GS} is varied from 0 to 6 V in steps of 1 V. The model parameters are L=10U W=20U VTO=2.5 KP=6.5E−3 RD=5 RS=2 RB=0 RG=0 RDS=1MEG.

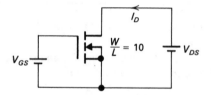

Figure P9.4

9.5. For Problem 9.4, plot the input characteristic if V_{GS} is varied from 0 to 6 V in steps of 0.1 V and $V_{DS} = 15$ V.

9.6. An inverter circuit is shown in Figure P9.6. For the input voltage as shown in Figure 9.21(b), plot the transient response of the output voltage from 0 to 80 μs in steps of 2 μs. If the input voltage is 5 V DC, calculate the voltage gain, the input resistance, and the output resistance. Print the small-signal parameters of the PMOS. The model parameters of the PMOS are VTO=−2.5 KP=4.5E−3 CBD=5PF CBS=2PF CGSO=1PF CGDO=1PF CGBO=1PF.

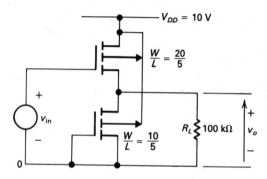

Figure P9.6

9.7. For the NMOS AND-logic circuit in Figure P9.7, plot the transient response of the output voltage from 0 to 100 μs in steps of 1 μs. The model parameters of the p-channel depletion-type MOSFETs are VTO=2 KP=4.5E-3 CBD=5PF CBS=2PF RD=5 RS=2 RB=0 RG=0 RDS=1MEG CGSO=1PF CGDO=1PF CGBO=1PF.

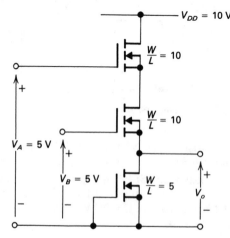

Figure P9.7

9.8. For the NMOS NAND-logic gate circuit in Figure P9.8, plot the transient response of the output voltage from 0 to 100 μs in steps of 1 μs. The model parameters of the PMOS are VTO=-2.5 KP=4.5E-3 CBD=5PF CBS=2PF RD=5 RS=2 RB=0 RG=0 RDS=1MEG CGSO=1PF CGDO=1PF CGBO=1PF. The model parameters of the NMOS are VTO=2.5 KP=4.5E-3 CBD=5PF CBS=2PF RD=5 RS=2 RB=0 RG=0 RDS=1MEG CGSO=1PF CGDO=1PF CGBO=1PF.

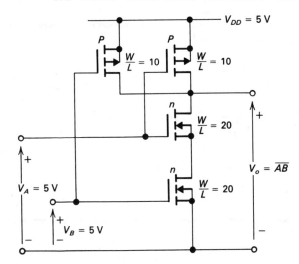

Figure P9.8

9.9. A MOSFET amplifier with active load is shown in Figure P9.9. Plot the magnitudes of the output voltage and the input current. The frequency is varied from 10 Hz to 100 MHz with a decade increment and 10 points per decade. The peak input voltage is 200 mV. The model parameters of the NMOS are VTO=2.5 KP=4.5E-2 CBD=5PF CBS=2PF RD=5 RS=2 RB=0 RG=0 RDS=1MEG CGSO=1PF CGDO=1PF CGBO=1PF. Print the details of the bias point and the small-signal parameters of the NMOS.

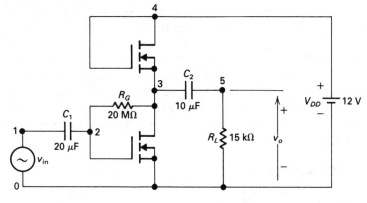

Figure P9.9

10

Op-Amp Circuits

10.1 INTRODUCTION

An **op-amp** may be modeled as a linear amplifier to simplify the design and analysis of op-amp circuits. The linear models give reasonable results, especially for determining the approximate design values of op-amp circuits. However, the simulation of the actual behavior of op-amps is required in many applications to obtain accurate response of the circuits. *PSpice* does not have any model for op-amps. However, an op-amp can be simulated from the circuit arrangement of the particular type of op-amp. The μ741 type of op-amps consists of 24 transistors, and it is beyond the capability of the students (or demo) version of *PSpice*. However, a macromodel, which is a simplified version of the op-amp and requires only two transistors, is quite accurate for many applications and can be simulated as a subcircuit or a library file. Without a complex model of an op-amp, the characteristic of op-amp circuits may be determined approximately by one of the following models:

DC linear models
AC linear model
Nonlinear macromodel

10.2 DC LINEAR MODELS

An op-amp may be modeled as a voltage-controlled voltage source, as shown in Figure 10.1(a). The input resistance is high, typically 2 MΩ, and the output resistance is very low, typically 75 Ω. For an ideal op-amp, the model in Figure 10.1(a) can be reduced to Figure 10.1(b). These models do not take into account the saturation effect and slew rate, which do exist in practical op-amps. The gain is also assumed to be independent of the frequency. But the gain of practical op-amps falls with the frequency. These simple models are normally suitable for DC or low-frequency applications.

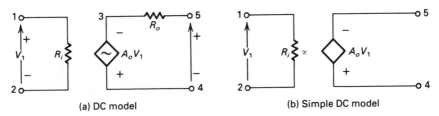

(a) DC model   (b) Simple DC model

Figure 10.1   DC linear models

10.3 AC LINEAR MODEL

An op-amp having one break frequency can be represented by a single capacitance, as shown in Figure 10.2. This is a high-frequency model of op-amps. If an op-amp has more than one break frequency, it can be represented by using as many capacitors as the number of breaks. R_i is the input resistance and R_o is the output resistance.

The dependent sources of the op-amp model in Figure 10.2 have a common node. Without this, *PSpice* will give an error message because there is no DC path from the nodes of the dependent current source. The common node could be either with the input stage or with the output stage. This model does not take into account the saturation effect and is suitable only if the op-amp operates within the linear region.

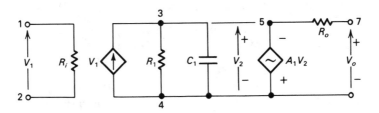

Figure 10.2   AC linear model with single break frequency

The output voltage can be expressed as

$$V_o = -A_1 V_2 = \frac{-A_1 R_1 V_i}{1 + R_1 C_1 s}$$

Substituting $s = j2\pi f$ yields

$$V_o = \frac{-A_1 R_1 V_i}{1 + j2\pi f R_1 C_1} = \frac{-A_o V_i}{1 + jf/f_b}$$

where
$$f_b = 1/(2\pi R_1 C_1) \text{ is called the } break \ frequency, \text{ in hertz}$$
$$A_o = A_1 R_1 \text{ is the } large\text{-}signal \text{ (or } dc) \ gain \text{ of the op-amp}$$

Thus, the open-loop voltage gain is

$$A(f) = \frac{V_o}{V_i} = -\frac{A_o}{1 + jf/f_b}$$

For $\mu 741$ op-amps, $f_b = 10$ Hz, $A_o = 2 \times 10^5$, $R_i = 2$ MΩ, and $R_o = 75$ Ω. Letting $R_1 = 10$ kΩ, $C_1 = 1/(2\pi \times 10 \times 10 \times 10^3) = 1.156$J9 μF.

10.4 NONLINEAR MACROMODEL

The circuit arrangement of the **op-amp macromodel** is shown in Figure 10.3 [1]. The macromodel can be used as a subcircuit with .SUBCKT command. However, if an op-amp is used in various circuits, it is convenient to have the macromodel as a library file, namely, NOM.LIB, and it is not required to type the statements of the macromodel in every circuit where the macromodel is employed. The library file that comes with the student version of the *PSpice* is NOM.LIB and contains the model for UA741 op-amp subcircuit. The professional version of the *PSpice* supports library files for many devices.

Note. The default library file for the latest version of PSpice is EVAL.LIB. Check your PSpice programs for the default library file.

The macromodel of the $\mu 741$ op-amp is simulated at the room temperature. The macromodel does not consider the effects of temperature. The library file, NOM.LIB, contains a model for nominal, not worst-case, devices. The macromodel consists of a subcircuit definition and a set of .MODEL statements.

The listing of the library file, NOM.LIB, is as follows.

```
* Library file "NOM.LIB" for u741 op-amp
* connections:   noninverting input
*                !   inverting input
*                !   !
*                !   !
*                !   !        positive power supply
*                !   !        !   negative power supply
*                !   !        !   !   output
*                !   !        !   !   !
.SUBCKT UA741    1   2        4   5   6
```

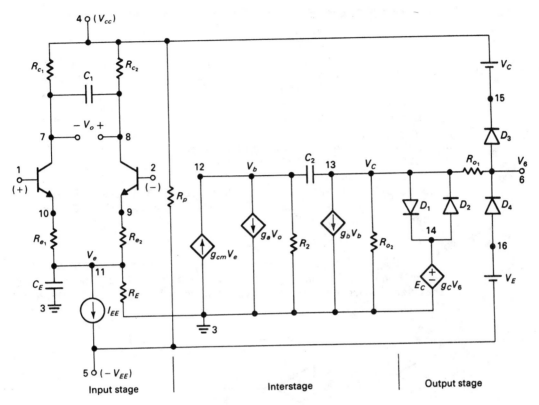

Figure 10.3   Circuit diagram of op-amp macromodel

```
*                      Vi+ Vi- Vp+ Vp- Vout
Q1    7    1    10    UA741QA
Q2    8    2    9     UA741QB
RC1   4    7    5.305165D+03
RC2   4    8    5.305165D+03
C1    7    8    5.459553D-12
RE1   10   11   2.151297D+03
RE2   9    11   2.151297D+03
IEE   11   5    1.666000D-05
CE    11   0    3.000000D-12
RE    11   0    1.200480D+07
GCM   3    12   11   3    5.960753D-09
GA    12   0    8    7    1.884955D-04
R2    12   0    1.000000D+05
C2    12   13   3.000000D-11
GB    13   0    12   3    2.357851D+02
RO2   13   0    4.500000D+01
D1    13   14   UA741DA
D2    14   13   UA741DA
```

```
EC    14   0    6    3    1.0
RO1   13   6    3.000000D+01
D3    6    15   UA741DB
VC    4    15   2.803238D+00
D4    16   6    UA741DB
VE    16   5    2.803238D+00
RP    4    5    18.16D+03
*  End of subcircuit definition
.ENDS
*  Models for diodes and transistors
.MODEL UA741DA D (IS=9.762287D-11)
.MODEL UA741DB D (IS=8.000000D-16)
.MODEL UA741QA NPN (IS=8.000000D-16 BF=9.166667D+01)
.MODEL UA741QB NPN (IS=8.309478D-16 BF=1.178571D+02)
*  End of library file
```

Example 10.1

An inverting amplifier is shown in Figure 10.4. The output is taken from node 5. Calculate and print the voltage gain, the input resistance, and the output resistance. The op-amp, which is modeled by the circuit in Figure 10.1, has $A_o = 2 \times 10^5$, $R_i = 2$ MΩ, and $R_o = 75$ Ω.

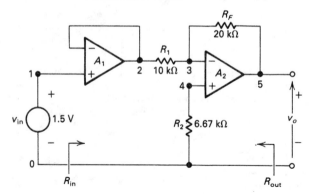

Figure 10.4 Inverting amplifier

Solution   The list of the circuit file is as follows.

```
Example 10.1     Inverting Amplifier
*   Input voltage is 1.5 V DC.
VIN  1  0   DC   1.5V
R1   2  3   10K
R2   4  0   6.67K
RF   3  5   20K
*  Calling subcircuit OPAMP
XA1  2  1   2   0   OPAMP
XA2  3  4   5   0   OPAMP
*  Subcircuit definition for OPAMP
.SUBCKT OPAMP 1  2  5  4
```

```
RIN 1   2   2MEG
RO  3   5   75
*  Voltage-controlled voltage source with a gain of 2E+5. The polarity of
*  the output voltage is taken into account by changing the location of the
*  controlling nodes.
EA  3   4   2   1   2E+5
*  End of subcircuit definition
.ENDS OPAMP
*   Transfer function analysis calculates and prints the DC gain,
*   the input resistance, and the out resistance.
.TF  V(5)  VIN
.END
```

```
****      SMALL-SIGNAL BIAS SOLUTION        TEMPERATURE =   27.000 DEG C
NODE    VOLTAGE      NODE    VOLTAGE      NODE    VOLTAGE      NODE    VOLTAGE
(   1)    1.5000  (    2)     1.5000  (    3) 15.11E-06  (    4) 50.21E-09
(   5)   -2.9999  (XA1.3)     1.5112  (XA2.3)   -3.0112
    VOLTAGE SOURCE CURRENTS
    NAME            CURRENT
    VIN           -3.778E-12
    TOTAL POWER DISSIPATION   5.67E-12  WATTS
```

```
****      SMALL-SIGNAL CHARACTERISTICS
     V(5)/VIN = -2.000E+00
     INPUT RESISTANCE AT VIN =   3.970E+11
     OUTPUT RESISTANCE AT V(5) =  1.132E-03
          JOB CONCLUDED
          TOTAL JOB TIME            2.42
```

Example 10.2

An integrator circuit is shown in Figure 10.5(a). For the input voltage as shown in Figure 10.5(b), plot the transient response of the output voltage for a duration of 0 to 4 ms in steps of 50 μs. The op-amp that is modeled by the circuit in Figure 10.2 has $R_i = 2$ MΩ, $R_o = 75$ Ω, $C_1 = 1.5619$ μF, and $R_1 = 10$ kΩ.

Solution The list of the circuit file is as follows.

```
Example 10.2    Integrator Circuit
*  The input voltage is represented by a piecewise linear waveform.
*  To avoid convergence problems due to a rapid change, the input
*  voltage is assumed to have a finite slope.
VIN  1   0   PWL (0   0 1NS -1V 1MS -1V  1.0001MS 1V 2MS 1V
+       2.0001MS -1V 3MS -1V  3.0001MS 1V  4MS 1V)
*  Transient analysis for 0 to 4 ms with 50-μs increment
.TRAN  50US  4MS
R1   1   2   2.5K
RF   2   4   1MEG
RX   3   0   2.5K
RL   4   0   100K
C1   2   4   0.1UF
```

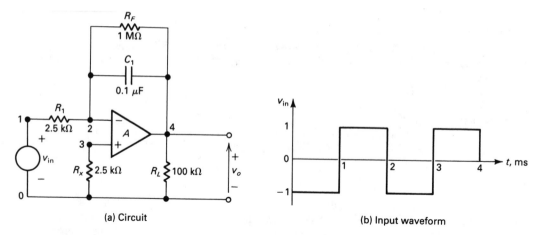

(a) Circuit   (b) Input waveform

Figure 10.5 Integrator circuit

```
*   Calling subcircuit OPAMP
XA1   2   3   4   0   OPAMP
*   Subcircuit definition for OPAMP
.SUBCKT OPAMP   1   2   7   4
RI   1   2   2.0E6
*   Voltage-controlled current source with a gain of 1
GB   4   3   1   2   1
R1   3   4   10K
C1   3   4   1.5619UF
*   Voltage-controlled voltage source with a gain of 20
EA   4   5   3   4   20
RO   5   7   75
*   End of subcircuit OPAMP
.ENDS
*   Plot the results of transient analysis
.PLOT   TRAN   V(4)   V(1)
.PLOT   AC   VM(4)   VP(4)
.PROBE
.END
```

The transient response for Example 10.2 is shown in Figure 10.6.

Example 10.3

A practical differentiator circuit is shown in Figure 10.7(a). For the input voltage as shown in Figure 10.7(b), plot the transient response of the output voltage for a duration of 0 to 4 ms in steps of 50 μs. The op-amp, which is modeled by the circuit in Figure 10.2, has R_i = 2 MΩ, R_o = 75 Ω, C_1 = 1.5619 μF, and R_1 = 10 kΩ.
Solution   The list of the circuit file is as follows.

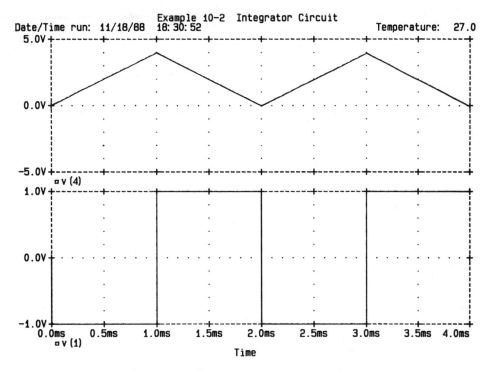

Figure 10.6 Transient response for Example 10.2

Example 10.3   Differentiator Circuit
```
*    The maximum number of points is changed to 410. The default
*    value is only 201.
.OPTIONS  NOPAGE  NOECHO LIMPTS=410
*  Input voltage is a piecewise linear waveform for transient analysis.
VIN 1  0  PWL (0  0  1MS  1  2MS  0  3MS  1  4MS  0)
```

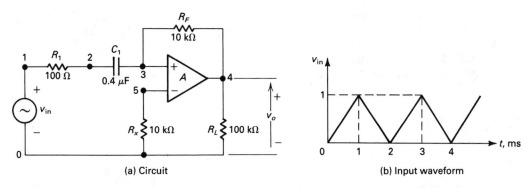

(a) Circuit

(b) Input waveform

Figure 10.7 Differentiator circuit

```
*  Transient analysis for 0 to 4 ms with 50 μs increment
.TRAN  10US  4MS
R1   1   2   100
RF   3   4   10K
RX   5   0   10K
RL   4   0   100K
C1   2   3   0.4UF
*  Calling op-amp OPAMP
XA1  3   5   4   0   OPAMP
*  Op-amp subcircuit definition
.SUBCKT  OPAMP  1   2   7   4
RI   1   2   2.0E6
*  Voltage-controlled current source with a gain of 1
GB   4   3   1   2   1
R1   3   4   10K
C1   3   4   1.5619UF
*  Voltage-controlled voltage source with a gain of 20
EA   4   5   3   4   20
RO   5   7   75
*  End of subcircuit OPAMP
.ENDS  OPAMP
*  Plot the results of transient analysis 4
.PLOT  TRAN  V(4)  V(1)
.PROBE
.END
```

The transient response for Example 10.3 is shown in Figure 10.8.

Example 10.4

A filter circuit is shown in Figure 10.9. Plot the frequency response of the output voltage. The frequency is varied from 10 HZ to 100 MHz with an increment of 1 decade and 10 points per decade. For the op-amp modeled by the circuit in Fig. 10.3, $R_i = 2$ MΩ, $R_o = 75$ Ω, $C_1 = 1.5619$ μF, and $R_1 = 10$ kΩ.

Solution The list of the circuit file is as follows.

```
Example 10.4    A Filter Circuit
*  AC analysis for 10 Hz to 100 MHz with a decade increment and
*  10 points per decade
.AC  DEC  10  10HZ  100MEGHZ
*  Input voltage is 1 V peak for ac analysis or frequency response.
VIN  1   0   AC  1
R1   1   2   20K
R2   2   4   20K
R3   3   0   10K
R4   1   5   10K
R5   4   5   10K
R6   6   7   100K
RL   7   0   100K
C1   2   4   0.01UF
```

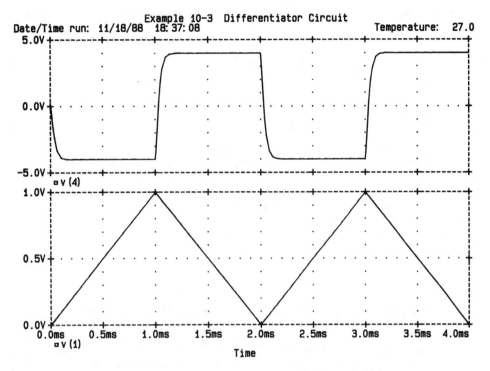

Figure 10.8   Transient response for Example 10.3

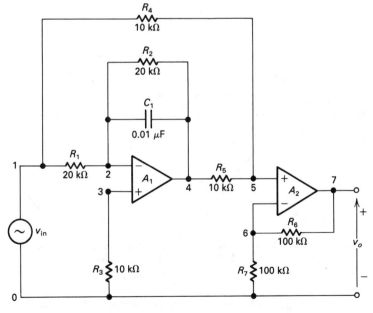

Figure 10.9   A filter circuit

```
*   Subcircuit call for OPAMP
XA1   2   3   4   0   OPAMP
XA2   5   6   7   0   OPAMP
*   Subcircuit definition for OPAMP
.SUBCKT   OPAMP   1   2   7   4
RI   1   2   2.0E6
*   Voltage-controlled current source with a gain of 1
GB   4   3   1   2   1
R2   3   4   10K
C2   3   4   1.5619UF
*   Voltage-controlled voltage source of gain 20
EA   4   5   3   4   20
RO   5   7   75
*   End of subcircuit definition
.ENDS   OPAMP
*   Plot the results of ac analysis
.PLOT   AC   VM(7)   VP(7)
.PROBE
.END
```

The frequency response for Example 10.4 is shown in Figure 10.10.

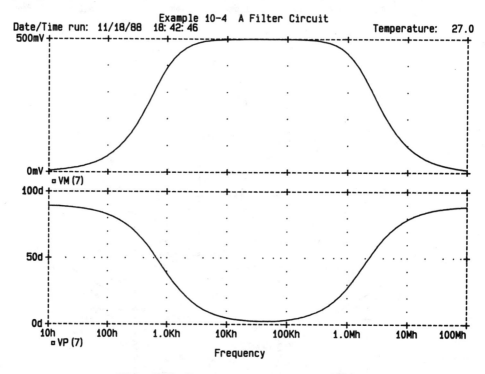

Figure 10.10 Frequency response for Example 10.4

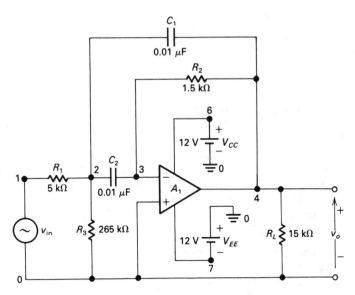

Figure 10.11 Band-pass active filter

Example 10.5

A band-pass active filter is shown in Figure 10.11. The op-amp can be modeled as a macromodel, as shown in Figure 10.3. The description of the UA741 macromodel is listed in library file NOM.LIB. Plot the frequency response if the frequency is varied from 100 Hz to 1 MHz with an increment of 1 decade and 10 points per decade. The peak input voltage is 1 V.

Solution The list of the circuit file is as follows.

```
Example 10.5    Band-Pass Active Filter
*  Input voltage of 1 V peak for frequency response
VIN   1   0   AC   1
*   AC analysis for 100 Hz to 1 MHz with a decade increment and 10
*   points per decade
.AC   DEC   10   100HZ   1MEGHZ
R1    1   2   5K
R2    3   4   1.5K
R3    2   0   265K
C1    2   4   0.01UF
C2    2   3   0.01UF
RL    4   0   15K
VCC   6   0   DC   12V
VEE   0   7   DC   12V
*   Subcircuit call for UA741
X1    0       3           6   7   4       UA741
*   Vi+   Vi-         Vp+   Vp-   Vout
*   Call library file NOM.LIB
.LIB   NOM.LIB
```

```
.PROBE
*  Plot the results of AC analysis: magnitude of voltage at node 4
.PLOT   AC   VM(4)
.PROBE
.END
```

The frequency response for Example 10.5 is shown in Figure 10.12.

Example 10.6

A free-running multivibrator circuit is shown in Figure 10.13. Plot the transient response of the output voltage for a duration of 0 to 4 ms in steps of 20 μs. The op-amp can be modeled as a macromodel as shown in Figure 10.3. The description of the UA741 macromodel is listed in library file NOM.LIB. Assume the initial voltage of the capacitor $C_1 = -5$ V.

Solution   The list of the circuit file is as follows.

```
Example 10.6      Free-Running Multivibrator
R1    1    0    100K
R2    1    2    100K
R3    2    3    10K
C1    3    0    0.1UF IC=-5V
VCC   6    0    DC   12V
```

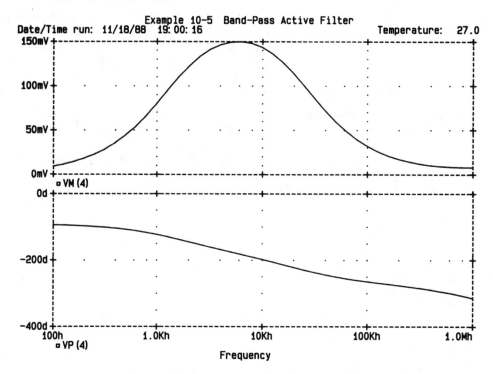

Figure 10.12   Frequency response for Example 10.5

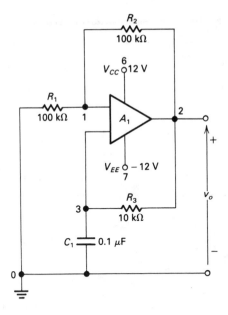

Figure 10.13   Free-running multivibrator

```
VEE   0  7 DC   12V
*  Subcircuit call for UA741
XA1   1      3           6   7    2      UA741
*     Vi+  Vi-         Vp+  Vp-  Vout
*  Call library file NOM.LIB
.LIB NOM.LIB
*  Transient analysis from 0 to 4 ms in steps of 20 µs
.TRAN  20US  4MS  UIC
.PROBE
.END
```

The transient response for Example 10.6 is shown in Figure 10.14.

Example 10.7

The circuit diagram of a differential amplifier with a transistor current source is shown in Figure 10.15. Calculate the DC voltage gain, the input resistance, and the output resistance. The input voltage is 0.1 V. The model parameters of the bipolar transistors are BF=50 RB=70 RC=40.

Solution   The list of the circuit file is as follows.

```
Example 10.7     Differential Amplifier
VCC   11    0     12V
VEE   0    10     12V
VIN   1    0  DC  0.25V
RC1   11    3     10K
RC2   11    5     10K
RE1   4    12     150
```

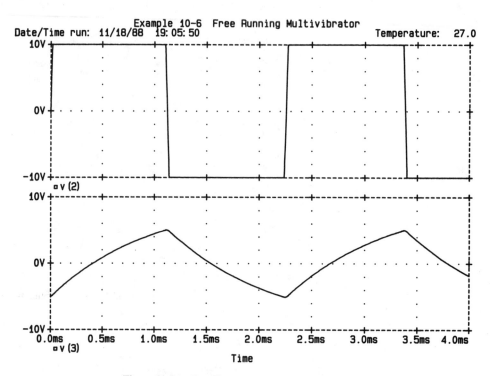

Figure 10.14 Transient response for Example 10.6

```
RE2    7   12   150
RS1    1    2   1.5K
RS2    6    0   1.5K
RX    11    8   20K
Q1     3    2    4   QN
Q2     5    6    7   QN
Q3    12    8    9   QN
Q4     9    9   10   QN
Q5     8    9   10   QN
*    DC transfer function analysis
.TF   V(3,5)   VIN
*     Model for NPN BJTs with model name QN
.MODEL QN NPN (BF=50 RB=70 RC=40)
.END
```

```
****    SMALL-SIGNAL BIAS SOLUTION      TEMPERATURE =   27.000 DEG C
 NODE    VOLTAGE       NODE    VOLTAGE      NODE    VOLTAGE       NODE    VOLTAGE
(    1)    .2500   (    2)     .2190   (    3)    1.6609   (    4)     -.5575
(    5)   11.3460  (    6)    -.0020   (    7)     -.7057  (    8)    -10.4430
(    9)  -11.2220  (   10)  -12.0000   (   11)   12.0000   (   12)     -.7157
```

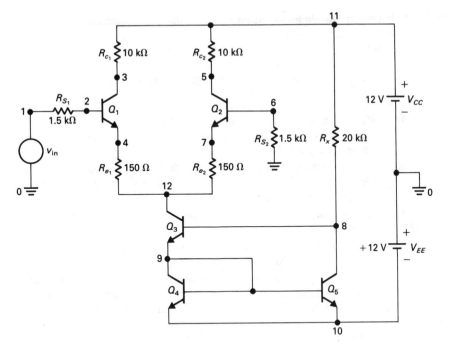

Figure 10.15   Differential amplifier

```
VOLTAGE SOURCE CURRENTS
NAME          CURRENT
VCC          -2.221E-03
VEE          -2.243E-03
VIN          -2.068E-05
TOTAL POWER DISSIPATION   5.36E-02   WATTS

****      SMALL-SIGNAL CHARACTERISTICS
    V(3,5)/VIN = -2.534E+01
    INPUT RESISTANCE AT VIN =  3.947E+04
    OUTPUT RESISTANCE AT V(3,5) =  2.000E+04
        JOB CONCLUDED
        TOTAL JOB TIME            4.01
```

REFERENCES

1. G. Boyle, B. Cohn, D. Pederson, and J. Solomon, "Macromodeling of integrated circuit operational amplifiers," *IEEE Journal of Solid-State Circuits,* Vol. SC-9, No. 6, December 1974, pp. 353–364.

2. I. Getreu, A. Hadiwidjaja, and J. Brinch, "An integrated-circuit comparator macro-model," *IEEE Journal of Solid-State Circuits,* Vol. SC-11, No. 6, December, 1976, pp. 826–833.

PROBLEMS

10.1. Plot the frequency response of the integrator in Figure 10.5 if the frequency is varied from 10 Hz to 100 kHz with a decade increment and 10 points per decade. The peak input voltage is 1 V.

10.2. Plot the frequency response of the differentiator in Figure 10.7 if the frequency is varied from 10 Hz to 100 kHz with a decade increment and 10 points per decade. The peak input voltage is 1 V.

10.3. Repeat Example 10.2 if the macromodel of the op-amp in Figure 10.3 is used. The supply voltages are $V_{CC} = 15$ V and $V_{EE} = -15$ V.

10.4. Repeat Example 10.3 if the macromodel of the op-amp in Figure 10.3 is used. The supply voltages are $V_{CC} = 15$ V and $V_{EE} = -15$ V.

10.5. A full-wave precision rectifier is shown in Figure P10.5. If the input voltage is $v_{in} = 0.1 \sin(2000\pi t)$, plot the transient response of the output voltage for a duration of 0 to 1 ms in steps of 10 μs. The op-amp can be modeled as a macromodel, as shown in Figure 10.3. The description of the UA741 macromodel is listed in library file NOM.LIB. Use the default values for the diode model. The supply voltages are

$$V_{CC} = 12 \text{ V and } V_{EE} = -12 \text{ V}$$

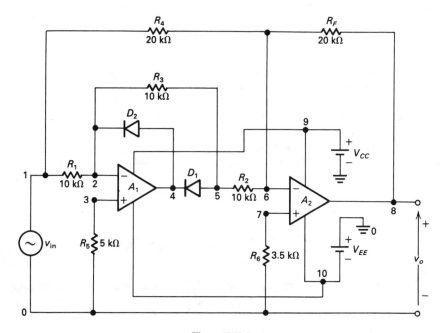

Figure P10.5

10.6. For Figure P10.5, plot the DC transfer characteristics. The input voltage is varied from -1 V to 1 V in steps of 0.01 V.

10.7. For Figure P10.7, plot the DC transfer characteristics. The input voltage is varied from -10 V to 10 V in steps of 0.1 V. The Zener voltages are $V_{Z1} = V_{Z2} = 6.3$ V. For the op-amp modeled by the circuit in Figure 10.1(a), $R_i = 2$ MΩ, $R_o = 75$ Ω, $C_1 = 1.5619$ μF, and $R_1 = 10$ kΩ.

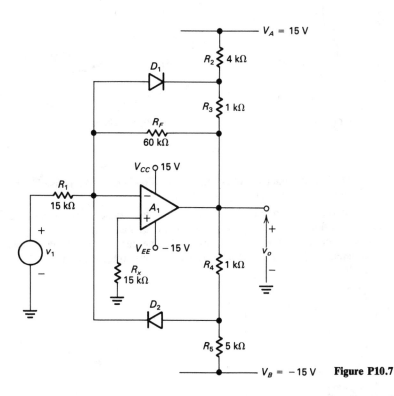

Figure P10.7

10.8. For Figure P10.8, plot the DC transfer function. The input voltage is varied from -10 V to 10 V in steps of 0.1 V. The zener voltages are $V_{Z1} = V_{Z2} = 6.3$ V. For the op-amp modeled by the circuit in Figure 10.1(a), $R_i = 2$ MΩ, $R_o = 75$ Ω, $C_1 = 1.5619$ μF, and $R_1 = 10$ kΩ.

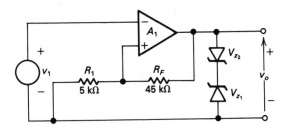

Figure P10.8

10.9. An integrator circuit is shown in Figure P10.9(a). For the input voltage as shown in Figure P10.9(b), calculate the slew rate of the amplifier by plotting the transient response of the output voltage for a duration of 0 to 200 μs in steps of 2 μs. For the op-amp modeled by the circuit in Figure 10.2, R_i = 2 MΩ, R_o = 75 Ω, C_1 = 1.5619 μF, and R_i = 10 kΩ.

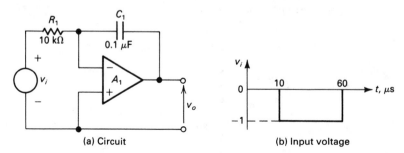

(a) Circuit   (b) Input voltage

Figure P10.9

10.10. Repeat Problem 10.9 if the macromodel of the op-amp in Figure 10.3 is used. The supply voltages are V_{CC} = 12 V and V_{EE} = -12 V.

10.11. A sine-wave oscillator is shown in Figure P10.11. Plot the transient response of the output voltage for a duration of 0 to 2 ms in steps of 0.1 ms. The op-amp can be modeled as a macromodel as shown in Figure 10.3. The description of the UA741 macromodel is listed in library file NOM.LIB. The supply voltages are V_{CC} = 15 V and V_{EE} = -15 V.

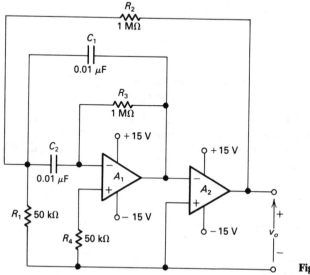

Figure P10.11

10.12. For the gyrator in Figure P10.12, plot the frequency response of the input impedance. The frequency is varied from 10 Hz to 10 MHz with a decade increment and
10 points per decade. For the op-amp modeled by the circuit in Figure 10.2, $R_i = 2$
$M\Omega$, $R_o = 75$ Ω, $C_1 = 1.5619$ μF, and $R_1 = 10$ $k\Omega$.

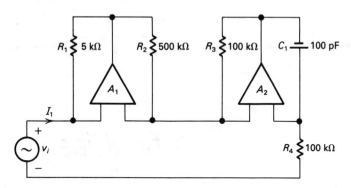

Figure P10.12

11

Difficulties

11.1 INTRODUCTION

An input file may not run for various reasons, and it is necessary to know what to do when the program does not work. To run a program successfully requires the knowledge of what would not work, why not, and how to fix the problem. There can be many reasons why a program does not work; this chapter covers the problems that are commonly encountered and their solutions. The problems could be due to one or more of the following causes:

Large circuits
Running multiple circuits
Large outputs
Long transient runs
Convergence
Analysis accuracy
Negative component values
Power switching circuits
Floating nodes

Nodes with fewer than two connections
Voltage source and inductor loops

11.2 LARGE CIRCUITS

The entire description of an input file must fit into RAM during the analyses.
However, none of the results of any analysis are stored into RAM. All results
(including intermediate results for the .PRINT and .PLOT statements) go to the
output file or one of the temporary files. Therefore, whether the run would fit into
RAM depends on how big the input file is.

The size of an input file can be found by using the ACCT option on the
.OPTIONS statement and looking at the MEMUSE number printed at the end of
runs that ran successfully. MEMUSE is the peak memory usage of that circuit.

If the circuit file does not fit, the possible remedies are

1. To break it up into pieces and run the pieces separately.
2. To reduce the amount of memory taken up by other resident software, e.g.,
   DOS and utilities. Total available memory can be checked by the DOS
   command: *CHKDSK*.
3. To buy more memory (up to 640K bytes, the most that PC-DOS recognizes).

11.3 RUNNING MULTIPLE CIRCUITS

A set of circuits may be run as a single job by putting all the circuits into one input
file. Each circuit begins with a title statement and ends with an .END command,
as usual. PSPICE1.EXE will read through all the circuits in the input file and then
process each one in sequence. The output file will contain the outputs from each
circuit in the same order as they appear in the input file. This technique is most
suitable for running a set of large circuits overnight, especially with the SPICE or
the professional version of *PSpice*. However, *Probe* cannot be used in this
situation because only the results of the last circuit will be available for graphical
output by *Probe*.

11.4 LARGE OUTPUTS

A large output file will be generated if an input file is run with several circuits, for
several temperatures, or with the sensitivity analysis. This will not be a problem
with a hard disk. For a PC with floppy disks, the diskette may be filled with the
output file. The best solution for this is

1. To direct the output to the printer instead of a file, or
2. To direct the output to an empty diskette instead of the one containing PSPICE1.EXE by assigning that the *PSpice* programs are on drive A and the input and output files on drive B. The command to run a circuit file would be: A:PSPICE B:EX2-1.CIR B:EX2-1.OUT.

11.5 LONG TRANSIENT RUNS

Long transient analysis runs can be avoided by choosing the appropriate limit options. The limits that affect the transient analysis are

1. The limit on the number of print steps in a run, LIMPTS,
2. The number of total iterations in a run, ITL5, and
3. The number of data points that *Probe* can handle.

The number of print steps in a run is limited to the value of the LIMPTS option. It has a default value of 0 (meaning no limit) but can be set to a positive value as high as 32000, e.g., .OPTIONS LIMPTS=6000. The number of print steps is simply the final analysis time divided by the print interval time (plus one). The size of the output file that is generated by *PSpice* can be limited in case of errors by the LIMPTS option.

The total number of iterations in a run is limited to the value of the ITL5 option. It has a default value of 5000, but it can be set at as high as 2×10^9, e.g., .OPTIONS ILT5=8000. The limit can be turned off by setting ITL5=0. This is the same as setting ITL5 to infinity and is often more convenient than setting it to a positive number.

Probe limits the data points to 16,000. This limit can be overcome by using the third parameter on the .TRAN statement to suppress part of the output at the beginning of the run. For a transient analysis from 0 to 10 ms in steps of 10 μs and printing output from 8 ms to 10 ms, the command would be

```
.TRAN 10US 10MS 8MS
```

Note. These options can be set from the "Change Options of Analysis" menu.

11.6 CONVERGENCE PROBLEMS

PSpice uses iterative algorithms. These algorithms start with a set of node voltages, and each iteration calculates a new set, which is expected to be closer to a solution of Kirchhoff's voltage and current laws. That is, an initial guess is used and the successive iterations are expected to converge to the solution. Convergence problems may occur in

DC sweep

Bias point calculation

Transient analysis

11.6.1 DC Sweep

If the iterations do not converge into a solution, the analysis fails. The DC sweep skips the remaining points in the sweep. The most common cause of failure of the DC sweep analysis is an attempt to analyze a circuit with regenerative feedback, such as Schmitt triggers. The DC sweep is not appropriate for calculating the hysteresis of such circuits because it is required to jump discontinuously from one solution to another at the crossover point.

To obtain hysteresis characteristics, it is advisable to use transient analysis with a piecewise linear (PWL) voltage source to generate a very slowly rising ramp. There is no CPU-time penalty for this because *PSpice* will adjust the internal time step to be large away from the crossover point and small in that region. A very slow ramp assures that the switching time of the circuit will not affect hysteresis levels. This is similar to changing the input voltage slowly until the circuit switches. With a PWL source in transient analysis, the hysteresis characteristics due to the upward and downward switching can be calculated.

Example 11.1

An emitter-coupled Schmitt trigger circuit is shown in Figure 11.1(a). Plot the hysteresis characteristic of the circuit from the results of the transient analysis. The input voltage, which is varied slowly from 1 V to 3 V and from 3 V to 1 V, is as shown in Figure 11.1(b). The model parameters of the transistors are IS=1E-16 BF=50 BR=0.1 RB=50 RC=10 TF=0.12NS TR=5NS CJE=0.4PF PE=0.8 ME=0.4 CJC=0.5PF PC=0.8 MC=0.333 CCS=1PF VA=50. Print the job statistical summary of the circuit.

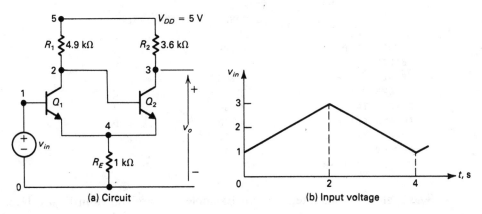

Figure 11.1   Schmitt trigger circuit

Solution   The input voltage is varied very slowly from 1 V to 3 V and from 3 V to 1 V, as shown in Figure 11.1(b). The list of the circuit file is as follows.

```
Example 11.1      Emitter-Coupled Trigger Circuit
*  Printing the accounts summary
.OPTIONS  ACCT
*  DC supply voltage of 5 V
VDD 5  0  DC  5
*  PWL waveform for transient analysis
VIN  1  0  PWL (0  1V  2  3V  4  1V)
R1   5  2  4.9K
R2   5  3  3.6K
RE   4  0  1K
*  Q1 and Q2 with model Qm
Q1   2  1  4  QM
Q2   3  2  4  QM
*  Model parameters for QM
.MODEL  QM NPN (IS=1E-16 BF=50 BR=0.1 RB=50 RC=10 TF=0.12NS TR=5NS
+  CJE=0.4PF PE=0.8 ME=0.4 CJC=0.5PF PC=0.8 MC=0.333 CCS=1PF VA=50)
*  Transient analysis from 0 to 4 s in steps of 0.01 s
.TRAN  0.01  4
.PROBE
.END
```

The job statistical summary obtained from the output file is as follows.

```
****      JOB STATISTICS SUMMARY
NUNODS  NCNODS  NUMNOD  NUMEL  DIODES    BJTS   JFETS   MFETS GASFETS
    6       6      10       7       0       2       0       0       0
NSTOP   NTTAR   NTTBR   NTTOV   IFILL    IOPS   PERSPA
   12      37      39      13       2      71   72.917
NUMTTP  NUMRTP  NUMNIT        MEMUSE
  285      55    1293         8906
                             SECONDS        ITERATIONS
       MATRIX SOLUTION        20.94             5
       MATRIX LOAD            59.29
       READIN                  1.54
       SETUP                    .05
       DC SWEEP               0.00              0
       BIAS POINT             5.54             77
       AC and NOISE           0.00              0
       TRANSIENT ANALYSIS   122.81           1293
       OUTPUT                 0.00
       TOTAL JOB TIME       124.90
```

The hysteresis characteristics for Example 11.1 are shown in Figure 11.2.

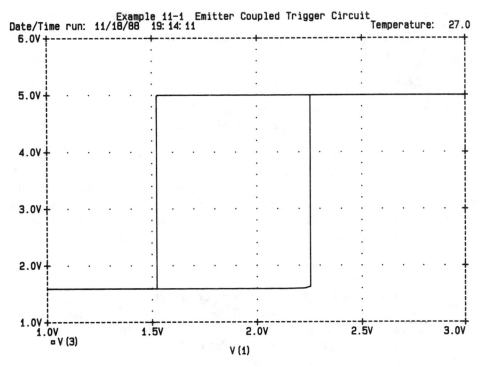

Figure 11.2 Hysteresis characteristics for Example 11.1

11.6.2 Bias Point

The failure of the bias point calculation prevents other analyses (e.g., AC analysis or sensitivity). The problems in calculating the bias point can be minimized by the .NODESET statement, e.g., .NODESET V(1)=0V. By giving *PSpice* "hints" in the form of initial guesses for node voltages, it starts out that much closer to the solution. A little judgment must be used in assigning appropriate node voltages.

It is rare to have a convergence problem in the bias point calculation. This is because *PSpice* contains an algorithm for automatically scaling the power supplies if it is having trouble finding a solution. This algorithm first tries to find a bias point with the power supplies at full scale. If there is no convergence, then the power supplies are cut back to one-fourth strength and the program tries again. If there is still no convergence, then the supplies are cut by another factor of 4 to one-sixteenth strength, and so on. Since at the power supplies of 0 V, the circuit will definitely have a solution with all nodes at 0 volts, the program will find a solution for some value of the supplies scaled far enough back. It then uses that solution to help it work its way back up to a solution with the power supplies at full strength. If this algorithm is in effect, a message such as

Power supplies cut back to 25%

(or some other percentage) appears on the screen while the program is calculating the bias point.

11.6.3 Transient Analysis

In case of failure due to convergence, the transient analysis skips the remaining time. The few remedies available for transient analysis, are

1. To change the relative accuracy, RELTOL, from 0.001 to 0.01, or
2. To set the iteration limits at any point during transient analysis by ITL4 option. Setting ITL4 = 50 (by statement .OPTIONS ITL5=50) will allow 50 iterations at each point. As a result of more iteration points, it will take longer simulation time. It is not recommended for circuits that do not have a convergence problem in transient analysis.

11.7 ANALYSIS ACCURACY

The accuracy of *PSpice*'s results is controlled by the parameters RELTOL, VNTOL, ABSTOL, and CHGTOL on the .OPTIONS statement. The most important of these is RELTOL, which controls the relative accuracy of all the voltage and currents which are calculated. The default value of RELTOL is 0.001 (0.1%).

VNTOL, ABSTOL, and CHGTOL set the best accuracy for the voltages, currents, and capacitor-charges/inductor-fluxes, respectively. If a voltage changes its sign and it gets close to zero, RELTOL will force *PSpice* to calculate more and more accurate values of that voltage because 0.1% of its value becomes a tighter and tighter tolerance. This would prevent *PSpice* from ever letting the voltage cross zero. To prevent this problem, VNTOL can limit all voltages' accuracy to a finite value; the default value is 1 μV. Similarly, ABSTOL and CHGTOL can limit the currents and charges (or fluxes), respectively.

The default values for the error tolerances in *PSpice* are the same as in the University of California, Berkeley, SPICE2. RELTOL = .001 (0.1%) is more accurate than that is necessary for many applications. The speed can be increased by setting RELTOL = 0.01 (1%), and this should increase the average speedup by a factor of 1.5.

Note. These options can be set from the "Change Options of Analysis" menu.

11.8 NEGATIVE COMPONENT VALUES

PSpice allows negative values for resistors, capacitors, and inductors. It should calculate a bias point or DC sweep for such a circuit. The .AC and .NOISE analyses can handle negative components. In the case of resistors, their noise contribution comes from their absolute values, and the components are not allowed to generate negative noise. However, negative components, especially negative capacitors and inductors, may cause instabilities in time, and the transient analysis may fail for a circuit with negative components.

Example 11.2

A circuit with negative resistances is shown in Figure 11.3. Calculate the voltage gain, the input resistance, and the output resistance.

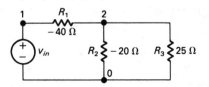

Figure 11.3   A circuit with negative resistances

Solution   The list of the circuit file is as follows.

```
Example 11.2     Circuit with Negative Components
*   DC input voltage of 1 V
VIN   1   0   DC   1V
*   Negative resistances
R1    1   2   -40
R2    2   0   -20
R3    2   0   25
*   DC transfer function analysis
.TF   V(2)   VIN
.END
```

```
****    SMALL-SIGNAL BIAS SOLUTION        TEMPERATURE =   27.000 DEG C
NODE    VOLTAGE      NODE   VOLTAGE     NODE   VOLTAGE     NODE    VOLTAGE
(    1)    1.0000  (    2)      .7143
    VOLTAGE SOURCE CURRENTS
    NAME          CURRENT
    VIN         7.143E-03
    TOTAL POWER DISSIPATION  -7.14E-03  WATTS

****    SMALL-SIGNAL CHARACTERISTICS
    V(2)/VIN =   7.143E-01
    INPUT RESISTANCE AT VIN = -1.400E+02
    OUTPUT RESISTANCE AT V(2) = -2.857E+01
        JOB CONCLUDED
        TOTAL JOB TIME            1.98
```

11.9 POWER-SWITCHING CIRCUITS

Running transient analysis on switching circuits can lead to long run times. *PSpice* must keep the internal time step short compared to the switching period, but the circuit's response extends generally over many switching cycles. This problem can be solved by transforming the switching circuit into an equivalent circuit without switching. The equivalent circuit represents a sort of "quasi-steady state" of the actual circuit and can accurately model the actual circuit's response as long the inputs do not change too fast.

Example 11.3

A single-phase bridge resonant inverter is shown in Figure 11.4(a). The transistors and diodes can be considered as switches whose on-state resistance is 10 mΩ and on-state voltage is 0.2 V. Plot the transient response of the capacitor voltage and the current through the load from 0 to 2 ms in steps of 10 μs. The output frequency of the inverter is $f_o = 4$ kHz.

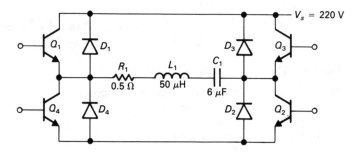

Figure 11.4   Single-phase bridge-resonant inverter

Solution   When transistors Q_1 and Q_2 are turned on, the voltage applied to the load will be V_s, and the resonant oscillation will continue for the whole resonant period first through Q_1 and Q_2 and then through diodes D_1 and D_2. When transistors Q_3 and Q_4 are turned on, the load voltage will be $-V_s$ and the oscillation will continue for another whole period first through Q_3 and Q_4 and then through diodes D_3 and D_4. The resonant period of a series RLC-circuit is approximately calculated as

$$\omega_r = \left(\frac{1}{LC} - \frac{R^2}{4L^2} \right)^{1/2}$$

For $L = L_1 = 50$ μH, $C = C_1 = 6$ μF, and $R = R_1 + R_{1(sat)} + R_{2(sat)} = 0.5 + 0.1 + 0.1 = 0.52$ Ω, $\omega_r = 57572.2$ rad/s and $f_r = \omega_r/2\pi = 9162.9$ Hz. The resonant period is $T_r = 1/f_r = 1/9162.9 = 109.1$ μs. The period of the output voltage is $T_o = 1/f_o = 1/4000 = 250$ μs.

The switching action of the inverter can be represented by two voltage-controlled switches, as shown in Figure 11.5(a). The switches are controlled by the voltages, as shown in Figure 11.5(b). The on-time of switches, which should be approximately equal to the resonant period of the output voltage, is assumed to be 112 μs. The switch S_2 is delayed by 115 μs to take into account overlap. The model parameters of the switches are RON=0.01 ROFF=10E+6 VON=0.001 VOFF=0.0.

The list of the circuit file is as follows.

```
Example 11.3    Full-Bridge Resonant Inverter
*    The controlling voltage for switch S1
V1   1   0   PULSE (0   220V   0   1US   1US   110US   250US)
```

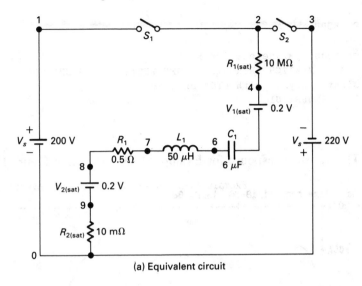

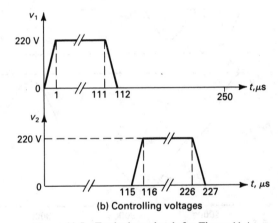

(a) Equivalent circuit

(b) Controlling voltages

Figure 11.5   Equivalent circuit for Figure 11.4

```
*    The controlling voltage for switch S2 with a delay time of 115 μs
V2   3   0   PULSE (0  -220V  115US  1US  1US  110US  250US)
*    Voltage-controlled switches with model SMOD
S1   1   2   1   0   SMOD
S2   2   3   0   3   SMOD
*    Switch model parameters for SMOD
.MODEL   SMOD   VSWITCH (RON=0.01   ROFF=10E+6   VON=0.001 VOFF=0.0)
RSAT1   2   4   10M
VSAT1   4   5   DC   0.2V
RSAT2   9   0   10M
VSAT2   8   9   DC   0.2V
*    Assuming an initial capacitor voltage of -250 V to reduce settling time
C1   5   6   6UF IC=-250V
```

```
L1   6   7   50UH
R1   7   8   0.5
*  Switch model parameters for SMOD
.MODEL   SMOD   VSWITCH (RON=0.01   ROFF=10E+6   VON=0.001 VOFF=0.0)
*  Transient analysis with UIC condition
.TRAN  2US   500US   UIC
.PROBE
.END
```

The transient response for Example 11.3 is shown in Figure 11.6.

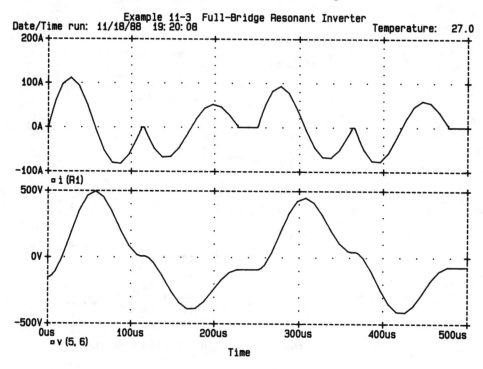

Figure 11.6   Transient responses for Example 11.3

11.10 FLOATING NODES

PSpice requires that there should be no floating nodes. If there are any, *PSpice* will indicate a read-in error in the screen, and the output file will contain a message similar to

```
ERROR: Node 15 is floating
```

This means that there is no DC path from node 15 to ground. A DC path is a path through resistors, inductors, diodes, and transistors. This is a very common problem, and it can occur in many circuits, as shown in Figure 11.7.

Node 4 in Figure 11.7(a) is floating and does not have a DC path. This problem can be avoided by connecting node 4 to node 0, as shown by dotted lines (or by connecting node 3 to node 2). A similar situation can occur in voltage-controlled and current-controlled sources, as shown in Figure 11.7(b) and (c). The model of op-amps as shown in Figure 11.7(d) has many floating nodes, which should be connected to provide DC paths to ground. For example, nodes 0, 3, and 5 could be connected together or, alternatively, nodes 1, 2, and 4 could be joined together.

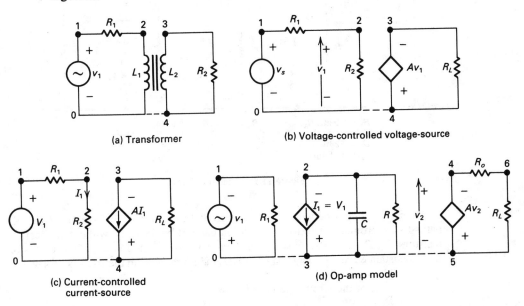

Figure 11.7   Typical circuits with floating nodes

The two sides of a capacitor have no DC path between them. If there are many capacitors in a circuit, as shown in Figure 11.8, nodes 3 and 5 do not have

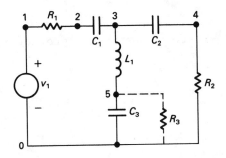

Figure 11.8   Typical circuit without DC path

DC paths. DC paths can be provided by connected a very large resistance R_3 (say 100 MΩ) across capacitor C_3, as shown by the dotted lines.

Example 11.4

A passive filter is shown in Figure 11.9. The output is taken from node 9. Plot the magnitude and phase of the output voltage separately against the frequency. The frequency should be varied from 100 HZ to 10 KHz in steps of 1 decade and 10 points per decade.

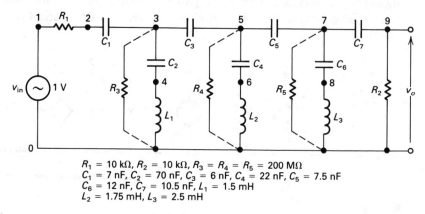

R_1 = 10 kΩ, R_2 = 10 kΩ, $R_3 = R_4 = R_5$ = 200 MΩ
C_1 = 7 nF, C_2 = 70 nF, C_3 = 6 nF, C_4 = 22 nF, C_5 = 7.5 nF
C_6 = 12 nF, C_7 = 10.5 nF, L_1 = 1.5 mH
L_2 = 1.75 mH, L_3 = 2.5 mH

Figure 11.9 A passive filter

Solution The nodes between C_1 and C_3, C_3 and C_5, and C_5 and C_7 do not have DC paths to the ground. Therefore, the circuit can not be analyzed without connecting resistors R_3, R_4, and R_5, as shown in Figure 11.9 by dotted lines. If the values of these resistance are very high, say 200 MΩ, their influence on the AC analysis would be negligible.

The list of the circuit file is as follows.

```
Example 11.4    A Passive Filter
*  AC analysis for 100 Hz to 10 kHz with a decade increment and
*  10 points per decade
.AC   DEC   10 100   10KHZ
*   Input voltage is 1 V peak for ac analysis or frequency response.
VIN   1   0   AC   1
R1    1   2   10K
R2    9   0   10K
*  Resistances R3, R4, and R5 are connected to provide dc paths
R3    3   0   200MEG
R4    5   0   200MEG
R5    7   0   200MEG
C1    2   3   7NF
C2    3   4   70NF
C3    3   5   6NF
C4    5   6   22NF
```

```
C5   5   7   7.5NF
C6   7   8   12NF
C7   7   9   10.5NF
L1   4   0   1.5MH
L2   6   0   1.75MH
L3   8   0   2.5MH
*  Plot the results of ac analysis for the magnitude of voltage
*  at node 9.
.PLOT   AC   VM(9)   VP(9)
.PLOT   AC   VP(9)
.PROBE
.END
```

The frequency response for Example 11.4 is shown in Figure 11.10.

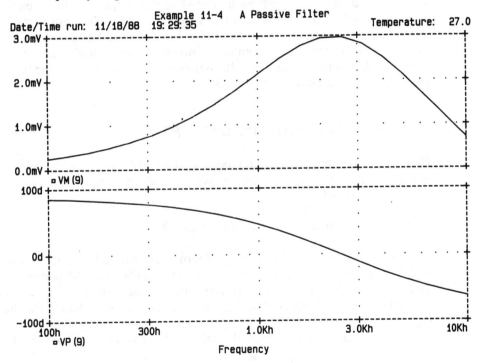

Figure 11.10   Frequency response for Example 11.4

11.11 NODES WITH LESS THAN TWO CONNECTIONS

PSpice requires that each and every node must be connected to at least two other nodes. Otherwise *PSpice* will give an error message, which is similar to

```
ERROR: Less than two connections at node 10
```

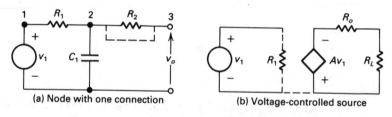

(a) Node with one connection   (b) Voltage-controlled source

Figure 11.11 Typical circuits with less than two connections at a node

This means that node 10 must have at least another connection. A typical situation is shown in Figure 11.11(a), where node 3 has only one connection. This problem can be solved by short-circuiting resistance R_2 as shown by the dotted lines.

An error message may be indicated in the output file for a circuit with voltage-controlled sources as shown in Figure 11.11(b). The input to the voltage-controlled source will not be considered to have connections during the check by *PSpice*. This is because the input draws no current and it has infinite impedance. A very high resistance (say, $R_i = 10$ GΩ) may be connected from the input to the ground, as shown by the dotted lines.

11.12 VOLTAGE SOURCE AND INDUCTOR LOOPS

PSpice requires that there are no loops with zero resistance. Otherwise, *PSpice* will indicate a read-in error on the screen, and the output file will contain a message similar to

```
ERROR:  Voltage loop involving V5
```

This means that the circuit has a loop of zero resistance components, one of which is V5. The zero resistance components in *PSpice* are: independent voltage sources (V), inductors (L), voltage-controlled voltage sources (E), and current-controlled voltage sources (H). Typical circuits with such loops are shown in Figure 11.12.

It does not matter whether the values of the voltage sources are 0 or not.

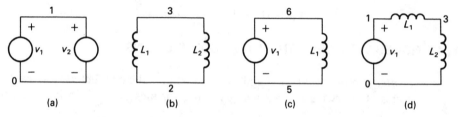

(a)   (b)   (c)   (d)

Figure 11.12 Typical circuits with zero resistance loops

Having a volt source of E in a zero resistance, the program will need to divide E by 0. But having E = 0 V, the program will need to divide 0 by 0, which also is impossible. It is, therefore, the presence of a zero resistance loop that is the problem, not the values of the voltage sources.

A simple solution is to add a series resistance to at least one component in the loop. The resistor's value should be small enough so that it does not disturb the operation of the circuit. However, the resistor's value should not be less than 1 $\mu\Omega$.

11.13 RUNNING *PSpice* FILE ON SPICE

PSpice will give essentially the same results as SPICE-2G from the University California, Berkeley (referred to as SPICE). There could be some small differences, especially for values crossing zero due to the corrections made for convergence problems. The semiconductor device models are the same as in SPICE.

There are a number of features of *PSpice* that are not available in SPICE:

1. Extended syntax for output variables (e.g., in .PRINT and .PLOT). SPICE allows only voltages of the form V(x) or V(x, y) and currents through voltage sources. Group delay is not available.
2. Extra devices:
   Gallium arsenide model
   Nonlinear magnetic (transformer) model
   Voltage- and current-controlled switch models
3. Optional models for resistors, capacitors, and inductors. The temperature coefficients for capacitors and inductors and exponential temperature coefficients for resistors.
4. The model parameters RG, RDS, L, W, and WD are not available in the MOSFET's .MODEL statement in SPICE.
5. Extensions to the DC sweep. SPICE restricts the sweep variable to be the value of an independent current or voltage source. SPICE does not allow sweeping of model parameters or temperature.
6. The .LIB and .INC statements.
7. SPICE requires the input (.CIR) file to be uppercase.

11.14 RUNNING SPICE FILE ON *PSpice*

PSpice will run any circuit, which SPICE-2G from the University of California, Berkeley (referred to as SPICE), will run with these exceptions:

1. Circuits that use .DISTO (small-signal distortion) analysis, which has errors in Berkeley SPICE. Also, the special distortion output variables (HD2,

DIM3, etc.) are not available. Instead of the .DISTO analysis, it is recommended to run a transient analysis and to look at the output spectrum with the Fourier transform mode of *Probe*. This technique shows the distortion (spectral) products for both small-signal and large-signal distortion.

2. The IN = option on the .WIDTH statement is not available. *PSpice* always reads the entire input file regardless of how long the input lines are.

3. Temperature coefficients for resistors must be put into a .MODEL statement instead of on the resistor statement. Similarly, the voltage coefficients for capacitors and the current coefficients for inductors are used in the .MODEL statements.

REFERENCES

1. *PSpice Manual*. Irvine, Calif.: MicroSim Corporation, 1988.
2. Wolfram Blume, "Computer Circuit Simulation," *BYTE,* Vol. 11, No. 7, July 1986, pp. 165.

PROBLEMS

11.1. For the inverter circuit in Figure P9.6, plot the hysteresis characteristics.

11.2. For the circuit in Figure P11.2, plot the hysteresis characteristics from the results of the transient analysis. The input voltage is varied slowly from -4 V to 4 V and from 4 V to -4 V. The op-amp can be modeled as a macromodel as shown in Figure 10.3. The description of the macromodel is listed in library file OPNOM.LIB. The supply voltages are $V_{CC} = 12$ V and $V_{EE} = -12$ V.

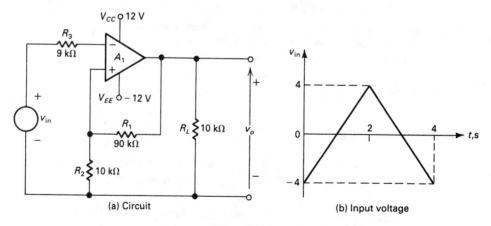

(a) Circuit   (b) Input voltage

Figure P11.2

Appendix A

Running PSpice on PCs

PSpice programs are available in high-density (1.2-megabyte) diskettes or normal (360-kilobyte) diskettes. The first step is to have the directory listing of the files on the program diskettes. The next step is to print and read the README.DOC file. It contains a brief description of the type of display and hard copy that are allowed by *PSpice* and *Probe*. It also contains the system requirements, and instructions on running *PSpice* programs.

PSpice will run on any IBM PC, the Macintosh II, or any compatible computer. The student version of *PSpice* does not require the coprocessor for running *Probe*. The display could be on monochrome or color graphics monitors. There is no requirement for special features of printers. The types of printers and display can be set by editing the PROBE.DEV file, or "Display/Printer Setup" menu.

The simulation of a circuit requires
Creating input files
Run command
DOS (disk operating system) commands

A.1 CREATING INPUT FILES

The latest version of PSpice has a built-in-editor with shell. The input files can also be created by text editors. The text editor that is always available is EDLIN. It comes with DOS and is described in the *DOS* user's guide. There are other editors, such as *Program Editor* (from WordPerfect Corporation). Word processing programs such as *WordStar, WordStar* 2000, and *Word* may also be used to create the input file. The word processor normally creates a file that is not a text file. It contains embedded characters to determine margins, paragraph boundaries, pages, and the like. However, most word processors have a command or mode to create a text file without these control characters. For example, *WordStar* 200 creates text files with UNIFORM format.

A.2 RUN COMMAND

PSpice is run from the menu, but *PSpice* can be run by typing

```
PSPICE  ⟨input file⟩  ⟨output file⟩
```

By default, the input file has the extension of .CIR and the output file has the extension of .OUT. The name of the output file defaults to the name of the input file. If the input file, EX2-1.CIR, is on the default drive, the following commands are equivalent:

```
PSPICE  EX2-1
PSPICE  EX2-1.CIR
PSPICE  EX2-1.CIR  EX2-1
PSPICE  EX2-1.CIR  EX2-1.OUT
```

The output file can be assigned to the printer that is connected to the PC by

```
PSPICE  EX1  PRN
```

The commands that will instruct *PSpice* about the location of the input file and the programs files will depend on the type of disk drives. There are two types of disks: fixed (or hard) disk and floppy disk, which is not fixed.

With a Fixed Disk.   Running *PSpice* with a fixed disk is straightforward. The PSPICE.BAT file must be in the default drive. Running *PSpice* will call the PSPICE.BAT file, which in turn will call the PSPICE1.EXE file and then the PROBE.EXE file, if required. PSPICE1.EXE creates one temporary file for storing intermediate results and deletes this temporary file when it finishes. If the circuit file EX2-1.CIR is in a diskette on drive A, the command for running *PSpice* is

```
PSPICE  A:EX2-1.CIR  A:EX2-1.OUT
```

Without a Fixed Disk.   The input file and the PSPICE.BAT file must be on Diskette 1 on drive A. The PSPICE.BAT file is searched by DOS for programs and commands. Running *PSpice* will cause PSPICE.BAT to call PSPICE1.EXE from drive B. PSPICE1.EXE creates one temporary file for storing intermediate results and deletes automatically the temporary file after writing to the output file when it finishes. The command for running the input file EX2-1.CIR is

```
PSPICE A:EX2-1.CIR
```

A.3 DOS COMMANDS

The DOS commands that are frequently used are the following:

To format a brand new diskette on drive A

```
FORMAT A:
```

To list the directory of a diskette on drive A

```
DIR   A:
```

To delete the file EX2-1.CIR on drive A

```
Delete A:EX2-1.CIR    (or  Erase A:EX2-1.CIR)
```

To copy the file EX2-1.CIR on drive A to the file EX2-2.CIR on drive B

```
COPY  A:EX2-1.CIR  B:EX2-2.CIR
```

To copy all the files on diskette in drive A to diskette on drive B

```
COPY  A:*.*  B:
```

To type the contents of file EX2-1.OUT on drive A

```
TYPE  A:EX2-1.OUT
```

To print the contents of the file EX2-1.CIR on drive A to the printer, first activate the printer by pressing Ctrl (Control) and Prtsc (Print Screen) keys together and then type

```
TYPE A:EX2-1.CIR
```

The printer can be deactivated by pressing Ctrl (Control) and Prtsc (Print Screen) keys again.

Appendix B

Noise Analysis

Noise is generated in electronic circuits. That is, an electronic circuit will have output even without any input signal. Noise can be classified as one of five types:

Thermal noise
Shot noise
Flicker noise
Burst noise
Avalanche noise

B.1 THERMAL NOISE

Thermal noise is generated in resistors due to random motion produced by the thermal agitation of electrons. This noise is dependent on temperature. The equivalent circuit for thermal noise in resistors is shown in Figure B.1. The mean square value of the noise generator is expressed as

$$V_t^2 = 4kTR\Delta f \qquad V^2$$

$$I_t^2 = 4kT\Delta f/R \qquad A^2$$

where

$$k = \text{Boltzmann's constant } (1.38 \times 10^{-23} \text{ J/K})$$

$$T = \text{absolute temperature, kelvin}$$

$$R = \text{resistance, ohms}$$

$$\Delta f = \text{noise bandwidth, hertz}$$

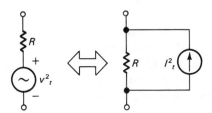

Figure B.1 Equivalent circuit for noise in resistors

B.2 SHOT NOISE

Shot noise is generated due to random fluctuations in the number of charged carriers when emitted from a surface or diffused from a junction. This noise is always associated with a direct current flow and is present in bipolar transistors. The mean square value of the noise current is expressed by

$$I_s^2 = 2qI_D \, \Delta f \qquad \text{A}^2$$

where

$$\Delta f = \text{noise bandwidth, hertz}$$

$$q = \text{electron charge } (1.6 \times 10^{-19} \text{ C})$$

$$I_D = \text{DC current, amps}$$

B.3 FLICKER NOISE

Flicker noise is generated due to surface imperfections resulting from the emission. This noise is associated with all active devices and some discrete passive elements such as carbon resistors. The mean square value of the noise current is expressed by

$$I_f^2 = K_f \frac{I_D^a}{f} \, \Delta f \qquad \text{A}^2$$

where

Δf = noise bandwidth, hertz

I_D = direct current

K_f = flicker constant for a particular device

a = flicker exponent constant in the range of 0.5 to 2

B.4 BURST NOISE

Burst noise is generated due to the presence of heavy metal ion contamination and is found in some integrated circuits and discrete transistors. The repetition rate of noise pulses is in the audio frequency range (a few kilohertz or less) and produces a "popping" sound when played through a speaker. This noise is also known as *popcorn noise*. The mean square value of the noise current is expressed as

$$I_b^2 = K_b \frac{I_D^c}{1 + (f/f_c)} \Delta f \qquad A^2$$

where

Δf = noise bandwidth, Hz

I_D = direct current

K_b = burst constant for a particular device

c = burst exponent constant

f_c = a particular frequency for a given noise

B.5 AVALANCHE NOISE

Avalanche noise is produced by Zener or avalanche breakdown in *pn*-junctions. The holes and electrons in the depletion region of a reverse-biased *pn*-junction acquire sufficient energy to create hole-electron pairs by collision. This process is cumulative, resulting in the production of a random series of large noise spikes. This noise is associated with direct current and is much greater than shot noise for the same current. Zener diodes are normally avoided in circuits requiring low noise.

B.6 NOISE IN DIODES

The equivalent circuit for noise in diodes is shown in Figure B.2. There are two generators. The voltage generator is due to thermal noise in the resistance of the silicon. The current source is due to the shot noise and flicker noise. The noise

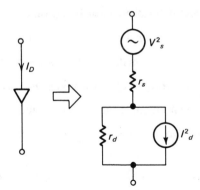

Figure B.2 Equivalent circuit for noise in diodes

voltage is given by

$$V_s^2 = 4kTr_s\Delta f \qquad \text{V}^2$$

$$I_d^2 = 2qI_D\Delta f + K_f \frac{I_D^a}{f}\Delta f \qquad \text{A}^2$$

where

I_D = forward diode current, amps

r_s = resistance of the silicon, ohms

K_f = flicker constant for a particular device

a = flicker exponent constant in the range of 0.5 to 2

B.7 NOISE IN BIPOLAR TRANSISTORS

The equivalent circuit for noise in bipolar transistors is shown in Figure B.3. The current generator in the collector is due to shot noise. The noise voltage generator in the base circuit is due to thermal noise in the base resistance. The current generator in the base circuit consists of shot noise, flicker noise, and burst noise. The noise is expressed by

$$V_b^2 = 4kTr_b\Delta f \qquad \text{V}^2$$

$$I_c^2 = 2qI_C\Delta f \qquad \text{A}^2$$

$$I_b^2 = 2qI_B\Delta f + K_f \frac{I_B^a}{f}\Delta f + K_b \frac{I_B^c}{1 + (f/f_c)}\Delta f \qquad \text{A}^2$$

where

I_B = base bias current, amps

I_C = collector bias current, amps

r_b = resistance at the transistor base, ohms

K_b = burst constant for a particular device

c = burst exponent constant in the range of 0.5 to 2

f_c = a particular frequency for a given noise

K_f = flicker constant for a particular device

a = flicker exponent constant in the range of 0.5 to 2

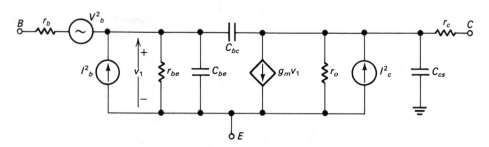

Figure B.3 Equivalent circuit for noise in bipolar transistors

B.8 NOISE IN FIELD-EFFECT TRANSISTORS

The equivalent circuit for noise in field-effect transistors (FETs) is shown in Figure B.4. The current generator in the gate is due to shot noise, which is very small. The current generator in the drain circuit consists of thermal and flicker noise. The noise current is expressed by

$$I_g^2 = 2qI_G\Delta f \qquad A^2$$

$$I_d^2 = 4kT \left(\frac{2}{3} g_m\right) \Delta f + K_f \frac{I_D^a}{f} \Delta f \qquad A^2$$

where

I_D = drain bias current, amps

I_G = gate leakage current, amps

g_m = transconductance at bias point, amps per volt

r_b = resistance of the transistor base, ohms

K_f = flicker constant for a particular device

a = flicker exponent constant in the range of 0.5 to 2

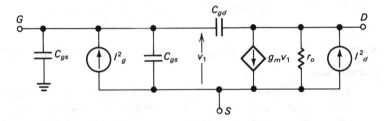

Figure B.4   Equivalent circuit for noise in field-effect transistors

B.9 EQUIVALENT INPUT NOISE

Each noise generator contributes to the output of a circuit. The effect of all the noise generators can be found by summing the mean square value of individual noise contributions. Once the total mean square noise output voltage is found, all the noises can be represented by an equivalent input noise at a desired source as shown in Figure B.5. This input noise is found by dividing the output voltage by

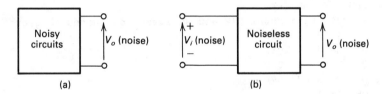

Figure B.5   Equivalent noise input

the gain. The gain is the output noise voltage with respect to a defined input. *PSpice* calculates the output noise and the equivalent input noise by the .NOISE command, which is covered in Section 6.8. It may be noted that *PSpice* calculates the noise in $V/\sqrt{Hz}$ or $A/\sqrt{Hz}$. Dividing the mean square value of the noise output voltage, $V_{o(noise)}$, by the noise bandwidth gives the output noise spectrum, V_o. That is,

$$V_o = \frac{V_{o(noise)}}{\Delta f} \qquad V/\sqrt{Hz}$$

Dividing the output noise spectrum by the gain yields the equivalent input noise spectrum as,

$$V_i = \frac{V_o}{G_v} \qquad V/\sqrt{Hz}$$

where G_v is the voltage gain with respect to the equivalent input source.

If the equivalent input is a current source, the equivalent input current noise spectrum becomes

$$I_i = \frac{V_o}{R_t} \quad V/\sqrt{Hz}$$

where R_t is the transresistance with respect to the equivalent input source.

Example B.1

For the inverter circuit in Figure 8.25, calculate and print the equivalent input and output noise. The frequency is varied from 1 Hz to 100 kHz with a decade increment and 1 point per decade. The input voltage is 1 V for AC analysis and 3.5 V for DC analysis.

Solution   The list of the circuit file is as follows.

```
Example B.1     TTL Inverter
*  Input voltage of 3.5 V for DC analysis and 1 V for AC analysis
VIN   1   0   DC   3.5V   AC   1V
VCC   13  0   5V
RS    1   2   50
RB1   13  3   4K
RC2   13  5   1.4K
RE2   6   0   1K
RC3   13  7   100
RB5   13  10  4K
*  BJTs with model QNP and substrate connected to ground by default
Q1    4   3   2   QNP
Q2    5   4   6   QNP
Q3    7   5   8   QNP
Q4    9   6   0   QNP
Q5    11  10  9   QNP
*  Diodes with model DIODE
D1    8   9   DIODE
D2    11  12  DIODE
D3    12  0   DIODE
*  Model of NPN transistors with model QNP
.MODEL QNP NPN (BF=50 RB=70 RC=40 TF=0.1NS TR=10NS VJC=0.85 VAF=50)
*  Diodes with model DIODE
.MODEL DIODE D (RS=40 TT=0.1NS)
*  AC sweep from 1Hz to 100 kHz with a decade increment
*  and 1 point per decade
.AC   DEC  1   1HZ   100KHZ
*  Noise analysis between output voltage, V(9) and input voltage, VIN
.NOISE   V(9)   VIN
*  Printing the results of noise analysis
.PRINT  NOISE  ONOISE INOISE
.END
```

The results of the noise analysis are given next.

```
****      SMALL-SIGNAL BIAS SOLUTION        TEMPERATURE =    27.000 DEG C
 NODE    VOLTAGE       NODE    VOLTAGE     NODE    VOLTAGE    NODE    VOLTAGE
(    1)    3.5000  (     2)     3.4724  (    3)    2.7564  (    4)    1.9134
(    5)    1.1702  (     6)     1.0205  (    7)    4.9998  (    8)     .5585
(    9)     .0635  (    10)      .9265  (   11)     .0816  (   12)     .0408
(   13)    5.0000
        VOLTAGE SOURCE CURRENTS
        NAME          CURRENT
        VIN          -5.524E-04
        VCC          -4.317E-03
        TOTAL POWER DISSIPATION    2.35E-02   WATTS

****      AC ANALYSIS                       TEMPERATURE =   27.000 DEG C
  FREQ        ONOISE       INOISE
   1.000E+00   8.146E-10   7.205E+02
   1.000E+01   8.146E-10   7.205E+02
   1.000E+02   8.146E-10   7.205E+02
   1.000E+03   8.146E-10   7.205E+02
   1.000E+04   8.146E-10   7.205E+02
   1.000E+05   8.146E-10   7.204E+02
          JOB CONCLUDED
          TOTAL JOB TIME        22.58
```

Appendix C

Nonlinear Magnetic

The nonlinear magnetic model uses MKS (metric) units. However, the results for *Probe* are converted to Gauss and Oersted and may be displayed using B(Kxx) and H(Kxx). The B-H curve can be drawn by a transient run with a slowly rising current through a test inductor and then by displaying B(Kxx) against H(Kxx).

Characterizing core materials may be done by trial by using *PSpice* and *Probe*. The procedures for setting parameters to obtain a particular characteristic are the following:

1. Set domain wall pinning constant, K = 0. The curve should be centered in the B-H loop, like a spine. The slope of the curve at H = 0 should be approximately equal to that when it crosses the *x*-axis at B = 0.
2. Set the magnetic saturation, MS = Bmax/0.01257.
3. Set the slope, ALPHA. Start with the mean field parameter, ALPHA = 0, and vary its values to get the desired slope of the curve. It may be necessary to change MS slightly to get the desired saturation value.
4. Change K to a nonzero value to create hysteresis. K effects the opening of the hysteresis loop.
5. Set C to obtain the initial permeability. *Probe* displays the permeability which is $\Delta B/\Delta H$. Since *Probe* calculates differences, not derivatives, so the

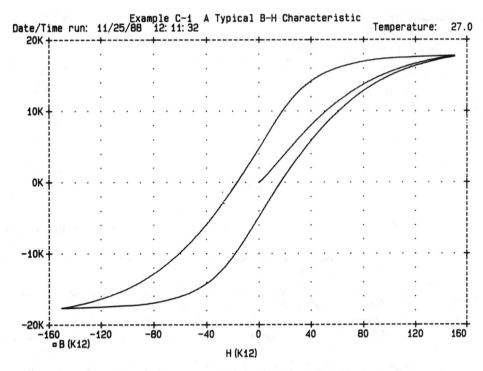

Figure C.1   A typical B-H characteristic

curves will not be smooth. The initial value of $\Delta B/\Delta H$ is the initial permeability.

Example C.1

The coupled inductors in Figure 5.4(a) are nonlinear. (See Figure C.1.) The parameters of the inductors are $L_1 = L_2 = 500$ turns, $k = 0.9999$. Plot the B-H characteristic of the core from the results of transient analysis if the input current is varied very slowly from 0 to -15 A, -15 A to 15 A, and from 15 A to -15 A. The load resistance of $R_L = 1$ KΩ is connected to the secondary of the transformer. The model parameters of the core are AREA=2.0 PATH=62.73 GAP=0.1 MS=1.6E+6 ALPHA= 1E$-$3 A=1E+3 C=0.5 K=1500.

Solution   The circuit file for the coupled inductors in Figure 5.4(a) would be as follows.

Example C.1   A Typical B-H Characteristic

```
*    PWL waveform for transient analysis
IN   1   0   PWL (0   0   1   -15   2   15   3   -15)
*    Inductors represent the number of turns
L1   1   0   500
L2   2   0   500
R2   2   0   1000
```

```
*   Coupled inductors with k = 0.9999 and model CMOD
K12  L1  L2  0.9999  CMOD
*   Model parameters for CMOD
.MODEL CMOD  CORE (AREA=2.0 PATH=62.73 GAP=0.1 MS=1.6E+6 ALPHA=1E-3 A=1E+3
+   C=0.5 K=1500)
*   Transient analysis from 0 to 3 s in steps of 0.03 s
.TRAN  0.05  3
.PROBE
.END
```

BIBLIOGRAPHY

1. Allen, Phillip E., *CMOS Analog Circuit Design*. New York: Holt, Rinehart and Winston, 1987.

2. Antognetti, Paolo, and Guiseppe Massobrio, *Semiconductor Device Modeling with SPICE*. New York: McGraw-Hill, 1988.

3. Bugnola, Dimitri S., *Computer Programs for Electronic Analysis and Design*. Reston, Va.: Reston Publishing Company, 1983.

4. Chua, Leon O., and Pen-Min Lin, *Computer-Aided Analysis of Electronic Circuits—Algorithms and Computational Techniques*. Englewood Cliffs, N. J.: Prentice Hall, 1975.

5. Ghandi, S. K., *Semiconductor Power Devices*. New York: John Wiley, 1977.

6. Gray, Paul R., and Robert G. Meyer, *Analysis and Design of Analog Integrated Circuits*. New York: John Wiley, 1984.

7. Grove, A. S., *Physics and Technology of Semiconductor Devices*. New York: John Wiley, 1967.

8. Hodges, D. A., and H. G. Jackson, *Analysis and Design of Digital Integrated Circuits*. New York: McGraw-Hill, 1988.

9. McCalla, William J., *Fundamentals of Computer-Aided Circuit Simulation*. Norwell, MA: Kluwer Academic, 1988.

10. MicroSim Corporation, *PSpice Manual*. Irvine, Calif.: MicroSim Corporation, 1988.

11. Nagel, Laurence, W., *SPICE2—A computer program to simulate semiconductor circuits*, Memorandum no. ERL-M520, May 1975, Electronics Research Laboratory, University of California, Berkeley.

12. Nashelsky, Louis, and Robert Boylestad, *BASIC for Electronics and Computer Technology*. Englewood Cliffs, N.J.: Prentice Hall, 1988.

13. Rashid, M. H., *Power Electronics—Circuits, Devices and Applications*. Englewood Cliffs, N.J.: Prentice Hall, 1988.

14. Spence, Robert, and John P. Burgess, *Circuit Analysis By Computer—from Algorithms to Package*. London: Prentice Hall International (UK), 1986.

15. Tuinenga, Paul W., *SPICE: A guide to circuit simulation and analysis using PSpice*, Englewood Cliffs, N.J.: Prentice Hall, 1988.

16. Ziel, Aldert van der, *Noise in Solid State Devices*. New York: John Wiley, 1986.

Index

ABSTOL 212
AC analysis 10
.AC statement 88
Active devices
  bipolar junction transistor (BJT) 119
  gallium-arsenide MESFET (GaAsFET) 177
  junction field effect transistor (JFET) 165
  models 101, 120
  MOS field effect transistor (MOSFET) 153
  semiconductor diode 101
Analysis 10
  AC 10
  DC 10
  transient 10
  types of 10
Avalanche noise 228

B-H curves 234
Bias-point 81
  regular 91
Bias-point analysis 81, 88
Bipolar junction transistor 119
  Ebers-Moll model of 121
  Gummel and Poon model of 119
  model parameters of 122
  model statement of 120
  small-signal model of 120
Burst noise 228

Capacitor
  model for 48
CHGTOL 212
Commands
  AC analysis 88
  DC analysis 81
  dot 67
  end of circuit 77
  Fourier analysis 95
  model 68
  noise analysis 89
  output 11, 71
  temperature 77
  transient response 91
Comment line 12
Continuation line 12
Convergence problems 208
Current-controlled current source 40
Current-controlled voltage source 40
Current-controlled switch 60
Current source
  current-controlled 40
  independent 37
  voltage-controlled 37, 39

DC analysis 81
DC bias-point 86
DC sensitivity 82
  .SENS statement 82

DC sweep 86
  .DC statement 86
Dependent sources 37
Device models 46
  .MODEL statement 102, 120, 154, 165
Devices
  symbols 21
Diode
  model parameters of 103
  statement of 104
  Zener 103
  semiconductor 101
DOS commands 225
Dot commands
  .AC 88
  .DC 86
  .END 77
  .ENDS 68
  .FOUR 96
  .IC 92
  .INC 70
  .LIB 70
  .MODEL 68
  .NOISE 90
  .NODESET 82
  .OP 81
  .OPTIONS 78
  .PLOT 71
  .PRINT 71
  .PROBE 72
  .SENS 82
  .SUBCKT 68
  .TF 84
  .TRAN 92
  .WIDTH 77

Ebers-Moll model 121
.ENDS (end of subcircuit definition) statement 68
.END statement 77
Elements
  capacitor 48
  circuit 7
  inductor 49
  initial conditions of 48
  magnetic 52
  modeling of 45
  models of 9
  passive 45
  resistor 47
  symbols of 8
  type name of 46
Exponential source 29

File
  circuit 2
    format of 11
  input 224
    creating 224
  output 12
    format of 12
Flicker noise 227

Fourier analysis 95
  .FOUR statement 96
Frequency response analysis 88
  .AC statement 88
Frequency sweep 89

Gallium-arsenide MESFET (GaAsFET) 177
  model parameters of 179
  model statement of 176
  small-signal model of 176
GMIN option 78
Group delay 24
Gummel-Poon model 119

.IC (initial transient conditions) statement 92
Imaginary party of complex values 24
.INC (include file) statement 70
Independent voltage source 36
Independent current source 37
Inductor coupling (transformer core) 52
Inductors
  model of 49
Initial conditions of elements 48
Initial transient conditions 93
INOISE 26
Input noise 26, 90, 231
ITL5 208

Jiles-Atherton magnetic model 54
Job statistics summary 79
Junction field effect transistor (JFET) 153
  Schichman and Hodges model of 153
  model parameters of 155
  model statement of 155
  small-signal model of 154

Large circuits 207
Large outputs 207
Library
  .LIB files 70
  .LIB statement 70
Library files
  CPNOM.LIB 188
  DNOM.LIB 188
  NNOM.LIB 188
  OPNOM.LIB 188, 197
  QNOM.LIB 188
LIMPTS 208
Linear frequency sweep 89
Logarithmic frequency sweep 89
Long transient runs 208
Loops
  inductor 220
  voltage source 220

Magnetics
  non-linear 54
Magnetic elements 52

Magnitude 24
MEMUSE 207
Metal Oxide field effect transistor (MOSFET) 153
  Schichman and Hodges model of 169
  model parameters of 167
  model statement of 165
  small-signal model of 166
Model parameters 46
Models
  BJT 120
  capacitor 48
  diode 102
  Gummel-Poon 119
  inductor 49
  JFET 155
  Jiles-Atherton magnetics 54
  MOSFET 165
  resistor 47
  Schichman-Hodges MOSFET
.MODEL statement 45
Mutual coupling, coefficient of 52
Mutual inductance 52

Names
  element 7
  model 46
Negative component values 212
Nodes 6
  circuit 6
  floating 216
  ground 6
.NODESET statement 82
Noise
  avalanche 228
  BJT 229
  Burst 228
  diode 228
  equivalent input 231
  Flicker 227
  JFET 230
  resistor 226
  shot 227
  thermal 226
Noise analysis 89
  .NOISE statement 90
Non-Linear inductor 54
Non-linear magnetics 234

ONOISE 26
op-amp macromodel 188
Op-amps 186
  DC linear models of 187
  AC linear model of 187
  nonlinear macromodel of 188
Operating Point 81
Options 78
  list of 78
.OP statement 81
.OPTIONS statement 78
Output
  current 22, 24
  voltage 21, 24

Passive elements 45
Phase angle 24
Piecewise linear source 31
.PLOT statement 71
Plus (+) sign 124
Polynomial source 33
.PRINT statement 71
Probe 4
Probe output 73
PROBE.DEV 77
PROBE.DAT 76
.PROBE statement 72
PSpice 2, 5
  limitations of 3
  running 221, 223
Pulse source 30

Real part of complex values 24
Resistors
  models of 47
Run command 224

Scaling element values 6
Semiconductor devices
  type names of 46
Semiconductor diode 101
Sensitivity analysis, DC 82
.SENS statement 82
Small-signal transfer function 84
Single-frequency FM source 31
Sinusoidal source 32
Small-signal DC analysis 86
Source(s) 9
  current-controlled current 40
  current-controlled voltage 40
  dependent 37
  exponential 29
  independent current 37
  independent voltage 36
  modeling 28
  piecewise linear 31
  polynomial 33
  pulse 30
  single-frequency frequency-modulation 31
  sinusoidal 32
  voltage-controlled current 39
  voltage-controlled voltage 37
SPICE 2
  types of 3
  running 221
SPICE2 2
SPICE3 2
Subcircuits 68
  .ENDS statement 68
  .SUBCKT statement 68
  X(subcircuit call) device 68
Suffixes
  scale 6
  units 6
Sweeping temperature 46

Switch
  current-controlled 60
  voltage-controlled 58
Symbols
  devices 21
  elements 8

Temperature
  operating 46
  sweeping 46
  .TEMP statement 46
.TF statement 84
Thermal noise 227
Transfer function
  analysis of 84
  circuit gain of 84
  input and output resistance of 84
  .TF statement 84
Transient response 91
  .TRAN statement 92
Transistor
  bipolar junction 119
  junction field effect 153
  MOS field effect 165
Transmission lines
  lossless 56
.TRAN statement 92

UIC (Use Initial Conditions) 93

Values
  element 6
    scaling 6
Variables
  AC analysis 24
  DC sweep 20
  noise analysis 26
  output 11
    defining 20
  transient analysis 20
VNTOL 212
Voltage-controlled current source 39
Voltage-controlled voltage source 37
Voltage-controlled switch 58
Voltage source
  current-controlled 40
  independent 36
  voltage-controlled 37

.WIDTH statement 77

X (subcircuit call) device 68

Zener diode 104